SUCCESSFUL BAKERY DESIGN

烘焙坊 II

(美)泰勒·鲁宾逊 编　李婵 译

辽宁科学技术出版社

CONTENTS
目录

Chapter 1 Shopfront and Branding Design
第一章 烘焙坊店面与品牌形象设计

- 006 **1.1 Shopfront Design** / **1.1 店面设计**
- 011 **1.2 Branding Identity** / **1.2 品牌形象**

Case Study / **案例解析**

- 014 A gift box along the street — 矗立在路边的礼品盒
- 015 A bakery shop hidden in the residential houses — 隐藏在居民区内的独特店面
- 016 A modern and eye-striking bakery along the street — 街对面的一家现代、引人注目的烘焙店面
- 018 A unique design in a quiet alley — 小巷弄里的独特设计
- 020 A distinctive renovation store at the square — 广场前别致的翻新店铺
- 022 A shopfront full of visibility and modernity — 良好的可见性，充满现代气息
- 024 Preserving the historic flavour — 保持历史风貌
- 026 Preserving the endearing aspects of Germany's confectionery culture in a district of cross-section classes — 在多阶层的街区内保留德国甜点文化的可爱之处
- 028 Visual branding inspired by the art of Lithuania and Belarus — 源于立陶宛和白俄罗斯的视觉形象设计
- 030 Communicating bakery's philosophy through identity design — 用品牌形象设计传达烘焙坊的生活哲学
- 032 Website design to connect delicious baked goods and the story behind them — 用网站设计将美味的烘焙食品及其背后的故事表出来……
- 034 Corporate identity to communicate the best homemade touch of the bakery — 用品牌形象设计表现自制烘焙食品的独特韵味
- 036 Identity design inspired by Buenos Aires' coat of arms — 灵感来自布宜诺斯艾利斯盾徽的品牌形象设计
- 038 Identity based on unique illustrations and brand character — 以独特的插画和品牌特色为出发点的品牌形象设

Chapter 2 Space Layout & Interior Design
第二章 空间布局与设计

- 042 **2.1 Space Layout** / **2.1 空间布局**
- 044 **2.2 Interior Design** / **2.2 室内设计**

Case Study / **案例解析**

- 050 Bulka Cafe and Bakery — 布尔卡咖啡烘焙坊
- 058 LOISIR — 洛伊斯烘焙坊
- 066 hey ju — "嘿！宇"烘焙工坊
- 072 The White Gift Box - Aimé Patisserie — Aimé 烘焙坊
- 078 Bakery Marie Antoinette — 玛丽皇后烘焙坊
- 084 Parken Bakery — 帕肯烘焙坊
- 090 The Cake — 蛋糕店

098	Royal Bakery		皇家烘焙坊
104	Patisserie Pan y Pasteles		西班牙马德里 Pan y Pasteles 甜品店
110	Vyta Santa Margherita Bakery		圣玛格丽塔烘焙坊
118	Zhengzhou Industrial Style Bakery		郑州工业风烘焙店
124	Kao Nekad Restaurant & Bakery		旧时光烘焙餐厅
130	Bakery MAISQUEPAN		梅斯科潘烘焙坊
134	Bartkowscy Bakery		巴特科沃斯基烘焙坊
140	BREAD, ESPRESSO&		BREAD, ESPRESSO& 面包咖啡店
146	BINARIO 11		拜纳里奥 11 号
154	DESIGNER'S VISION - Colour and forms serve as 'input' to attract customers walking in		设计构想——以色彩与形态来吸引顾客走进来

Chapter 3 Display Design

第三章 陈列设计

162	**3.1 Display Furniture and Methods**		**3.1 陈列家具与方式**
164	**3.2 Lighting for Display**		**3.2 陈列照明**
	Case Study		**案例解析**
166	Bread Table		日本岐阜面包桌烘焙坊
170	Pâtisserie À La Folie		疯狂点心店
176	Les Bébés Cafe & Bar		贝贝西点咖啡
182	Boulanger Kaiti		日本福冈卡提面包店
188	midi a midi		迷迪面包店
194	T by Luxbite		T 甜点屋
200	Style Bakery		风尚面包店
206	So Milaky		索·米拉其面包店
212	Serrajòrdia		塞拉约迪亚面包店
218	Chocolateria Brescó		布雷西科巧克力烘焙坊
224	IL LAGO Bakery & Wine Shop in MVL Hotel Kintex		韩国国际会展中心 MVL 酒店湖泊烘焙坊与酒廊
230	Gail's Bakery, Chelsea		英国伦敦切尔西盖尔面包店
238	**Index**		**索引**
	Appendix		**附录**

Chapter 1 Shopfront and Branding Design

第一章 烘焙坊店面与品牌形象设计

1.1 Shopfront Design
1.1 店面设计
1.2 Branding Identity
1.2 品牌形象

Case Study
案例解析

A gift box along the street
矗立在路边的礼品盒

A bakery shop hidden in the residential houses
隐藏在居民区内的独特店面

A modern and eye-striking bakery along the street
街对面的一家现代、引人注目的烘焙店面

A unique design in a quiet alley
小巷弄里的独特设计

A distinctive renovation store at the square
广场前别致的翻新店铺

A shopfront full of visibility and modernity
良好的可见性，充满现代气息

Preserving the historic flavour
保持历史风貌

Preserving the endearing aspects of Germany's confectionery culture in a district of cross-section classes
在多阶层的街区内保留德国甜点文化的可爱之处

Visual branding inspired by the art of Lithuania and Belarus
源于立陶宛和白俄罗斯的视觉形象设计

Communicating bakery's philosophy through identity design
用品牌形象设计传达烘焙坊的生活哲学

Website design to connect delicious baked goods and the story behind them
用网站设计将美味的烘焙食品及其背后的故事表现出来……

Corporate identity to communicate the best homemade touch of the bakery
用品牌形象设计表现自制烘焙食品的独特韵味

Identity design inspired by Buenos Aires' coat of arms
灵感来自布宜诺斯艾利斯盾徽的品牌形象设计

Identity based on unique illustrations and brand character
以独特的插画和品牌特色为出发点的品牌形象设计

Chapter 1
Shopfront and Branding Design

第一章
烘焙坊店面与品牌形象设计

1.1

With the improvement of people's living standards nowadays, the baking industry develops in an amazing way; a variety of bakeries spring out around the streets in different countries. Consumption behaviour often starts at the moment a customer steps into the bakery, and thus how to attract customers in is of great importance. Shopfront and branding design are two aspects that should be considered first.

1.1 Shopfront Design

A shopfront of a bakery with a bad design can be the consequence of cheap materials, bad workmanship and lack of thought but, more often, bad design is due to a lack of understanding of the value and importance of the elements that form original shopfronts to make them a visually cohesive part of the building. Good design, whether modern or traditional, recognises the importance of various elements of the shopfront, and integrates the aspirations of the shop owner without detrimentally affecting the building or its context. Good design can enhance and make positive contributions to the building, street scene and bakery operation. (See figure 1.1)

Preparation Work

Some aspects must be taken into consideration before the design of a bakery shopfront, including the street scene and local context, the rest of the building where the bakery is located, and the corporate identity of the bakery itself. As in regard of each aspect, the following questions need to be settled:

The street scene and local context

·What is the rhythm of the street elevation in which the shopfront will be located;

· Is there a consistent pattern to the shopfronts of adjoining buildings;

· What are the materials and colours used in the local buildings;

随着当今生活水平的提高，烘焙行业正在飞速发展，各类烘焙坊不断涌现在世界各国的街巷。消费行为往往始于顾客踏入店内的那一刻，因此，如何将顾客吸引进店就变得十分重要。店面和品牌形象设计是首先要考虑的两个方面。

1.1 店面设计

烘焙坊的店面如果设计得不好，原因可能是廉价的材料、施工技术不佳或者是设计上欠考虑，但是更常见的原因是：缺乏对店面构成元素的价值和重要性的理解，正是这些元素构成一家烘焙坊独特的店面形象，并且使其在视觉上跟整栋建筑融为一体。好的设计，不论是现代风格还是传统风格，首先要确立店面上各个构成元素的重要性，然后要考虑店主的愿望，同时不能对所在建筑物或者周围环境造成破坏。好的设计会对整条街道、整栋建筑以及烘焙坊的经营起到积极作用。（见图1.1）

准备工作

着手进行烘焙坊的店面设计之前，首先要考虑几个方面的问题，包括：街道和周围环境、店铺所在建筑的其余部分以及烘焙坊自身的品牌形象。具体到每个方面，需要解决以下问题：

街道和周围环境：

· 店面所在街道上，建筑物立面形成何种节奏韵律

· 毗邻建筑物的店面是否有统一的模式

· 周围建筑用的是什么材料和颜色

· 如何让新的店面融入街道环境

TYPICAL FORM OF A VICTORIAN OR EDWARDIAN SHOPFRONT

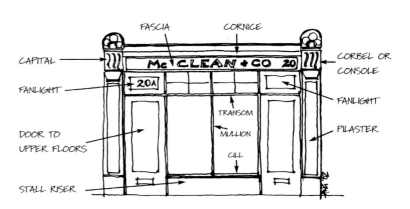

BASIC ELEMENTS OF A TRADITIONAL SHOPFRONT DESIGN

1.2

· How would a new shopfront fit in with the street scene.

The rest of the building

· What are the qualities and proportions of the rest of the building;

· Is there a particular architectural style;

· What materials are used;

· How well does the shopfront fit with the rest of the building;

The corporate identity of the bakery itself

·How to incorporate the brand identity into the shopfront;

·How to highlight the brand identity on the shopfront;

·How to express the brand culture through shopfront.

Design Points

After being clear of the above questions, it comes into design phase. The following three points are taken into account:

The form of the shopfront

·The proportion of scale and segmentation in elevation

As for the bakery located on the ground floor of a building, the existing architectural framework (vertical, horizontal rhythms, scale and proportions), determines the shopfront design. It normally comprises pilasters, with architectural details such as capital and plinth, a corbel or a console bracket, and an entablature with a frieze or fascia, and is terminated by a cornice. As for a detached bakery structure, the rhythm and proportions of shopfront needs to be carefully considered by its designer. (See figure 1.2)

建筑其余部分：

· 建筑其余部分的品质和比例尺度如何

· 有无某种建筑风格

· 用了什么材料

· 新店面与建筑其余部分是否能和谐搭配

烘焙坊自身品牌形象：

· 如何将品牌形象融入店面

· 如何在店面上突出品牌形象

· 如何通过店面来表现品牌文化

设计要点

清楚上述问题后，就进入到设计阶段。主要涉及以下三个问题。

店面的形态

· 立面的比例划分

对于占据建筑一楼位置的烘焙坊来说，建筑的框架结构（垂直和水平方向的韵律、比例的划分）决定了店面的设计。店面的结构一般包括壁柱（上面有柱顶和柱基等细部结构）、枕梁或支架、屋檐（安装招牌牌匾）以及檐口（飞檐）。如果一家烘焙坊是一座独栋建筑物，那么店面的韵律和比例划分就需要由设计师来谨慎决定。（见图1.2）

店面的构成也是建筑框架结构的一部分，主要包括门、窗、台阶，以及划分窗玻璃的直棂（如果有的话）。一个店面，如果各个构成部分的

Chapter 1
Shopfront and Branding Design

第一章
烘焙坊店面与品牌形象设计

1.3

1.4

The shopfront is the screen or panel that fills the space defined by architectural framework. It includes door, window, stallriser as well as any glazing bars (transoms, mullions) that might be present. A well proportioned shopfront can create surprisingly effect.

·The material selection and colour scheme

Materials should be selected to harmonise with the character of the building concerned. As a general principle, the type and number of materials used should be kept to a minimum and should be durable and easy to maintain.

Traditional materials such as painted timber, glass, steel, stone, glazed tiles are still the most commonly used for good bakery shopfronts and will appear in most better-quality designs. The use of timber can create a natural and cosy appearance, while the use of steel can also result in elegant simple modern designs. For example, in the shopfront of BREAD, ESPRESSO& in Taipei, red cedar is extensively used. In the shopfront of Bakery MAISQUEPAN in Galicia, the designer selects stainless steel to construct the shopfront. From a general view, most bakeries favour glass in shopfront for its unique characteristics of sold products. (See figure 1.3~1.4)

Here, aluminium is a modern material that comes in a variety of powder-coated finishes which may be acceptable as a cheap alternative to steel where a contemporary design is appropriate. But, natural or anodised aluminium weathers badly and is not acceptable for shopfront frames, doors or windows.

Colour for a bakery shopfront needs to be alluring like the baked products. At the same time, it requires to be simple to boast the heroes behind. White and transparent bakery shopfronts are commonly seen. For example, the shopfront of Style Bakery in Gunma, Japan is completely transparent with the inside being clearly shown. The White Gift Box - Aimé Patisserie in Shanghai is fully white in shopfront to give a neat while stylish feeling. (See figure 1.5~1.7)

比例划分得当，能带来令人惊喜的效果。

·材料选择与色彩搭配

材料的选择应该与所在建筑的特点相符。总的原则是：材料的种类和数量应该尽量减少，而且材料应该经久耐用，便于维护。

传统材料，如上漆木材、玻璃、钢、石材、釉面砖等，仍是烘焙坊店面设计最常用的材料，在许多质量上乘的设计中都可以见到。使用木材能营造出一种自然、舒适的感觉，而钢材则常用于现代、简约风格的设计。比如说，台北"BREAD, ESPRESSO&"烘焙坊的店面，大量使用了红松木。而在 MAISQUEPAN 烘焙坊店面上，设计师就用了不锈钢。但一般来说，大部分烘焙坊的店面都偏爱使用玻璃，以便展示店内产品的特色。（见图1.3、图1.4）

铝材也是一种现代材料，表面可以加各种覆层，可以作为钢材的廉价替代品，也符合现代的风格。但是，天然铝材或者阳极氧化铝不耐侵蚀，不可用作店面的框架或门窗。

烘焙坊店面的色彩应该像店内的产品一样能吸引人们的目光。同时，店面的色彩还要做到简洁，因为真正的主角是后面待售的产品。白色或者透明的烘焙坊店面很常见。比如说，日本群马县的风尚烘焙坊（Style Bakery），店面是完全透明的，室内通透可见。"白色礼盒–Aimé Patisserie"上海旗舰店店面是纯白色，简约、时尚、现代。（见图1.5～图1.7）

色彩也是用来强调设计中的重要元素，借以凸显某些方面或者某些细节，比如脚线或文字。另外，针对有视力障碍的人群，色彩也能突出

1.5

1.6

Colour can also be used to emphasise important elements of the design to reinforce certain aspects and to pick up details, such as mouldings and lettering. It is helpful to emphasise the location of bakery entrances for people with a visual impairment. This can be done through use of colour and textural contrast, on the vertical plane between entrance and the rest of the shopfront and underfoot, by emphasising the change from pavement to the bakery.

The window and doorway

The scale and proportion and profile of window frames, glazing bars and door locations should be derived from the characteristics of the street and the architectural style of the upper floors.

1.7

Entrance doors are normally best located centrally or adjacent to either pilaster. The location of entrance doors needs to respect the established rhythm within the street scene.

Windows and doors, through the appropriate use of colour, interesting shapes and proportions, quality materials and lively window displays can add visual interest to the street scene and produce a distinctive individual shopfront image. For example, in the design of Boulanger Kaiti in Fukuoka, Japan, large area of white is selected to cover the window while the doorway on one side is mainly in wood colour, which forms an interesting contrast. (See figure 1.8)

A variation in the plane, by recessing doors or curving windows, can add to visual interest. Deeply recessed windows or completely open frontages are mostly unacceptable in visual and functional terms.

烘焙坊入口的位置，可以在店面立面上让入口和其余部分在色彩和质地上形成鲜明的对比，在地面铺装上也可以突出店内外的差别。

门窗：

门窗的位置以及比例的划分应该根据街道的特点以及所在建筑物的风格来决定。

入口大门一般最好设置在中间，或者紧靠壁柱。入口的位置应该根据街道立面既定的韵律感来布置。

除了适当的色彩、形态和比例，品质上乘的材料和生机勃勃的橱窗展示，不但能营造出别具一格的店面形象，也能为整个街道的形象添彩。比如说，日本福冈市的"Boulanger Kaiti"烘焙坊，店面橱窗上采用大面积的白色，门廊则

Chapter 1
Shopfront and Branding Design

第一章
烘焙坊店面与品牌形象设计

Fascias and Signs

Fascias are perhaps the most dominant feature of the shopfront. An attractive fascia is also very important for a bakery shopfront. They play a dual role in both communicating the basic information such as name, trade and number of the bakery and forming an important design element in the 'framework' of the shopfront. Shop signs are similarly important; depending on their position, design and numbers they can either 'clutter' or add a delightful richness and variety to the street scene. (See figure 1.9)

In the design of fascia and signs, the following suggestions are given:

·The scale of the fascia should be in harmony with the other elements of the building where the bakery is located;

·Fascias or projecting signs or advertisements should not obscure the existing windows and architectural details;

·The use of large areas of acrylic or other shiny materials in fascias is better avoided;

·Box fascias, usually of plastic and/or metal, are not commonly seen in modern shopfront design;

·Large areas of glazing can be a useful location for signs, which can be painted or etched onto the internal surface of the windows. But remember to avoid excessive signage to make the windows not so cluttered;

·Projecting or hanging signs must be simple to communicate necessary information ;

·Keep the lettering legible and the style in harmony with the other elements of the building;

·Good contrast and simple lettering will make signage more legible.

是木材的颜色，形成有趣的对比。（见图1.8）

立面上也可以制造一些变化，比如让入口凹进，或者采用弧形窗，能够增加视觉效果。但是，大幅度凹进的橱窗或者彻底开放式的门脸，一般来说，不论从视觉效果还是功能上，都是不可取的。

招牌和标识：

店铺的招牌可能是店面最重要的部分了。有吸引力的招牌对一家烘焙坊的店面来说也很重要。招牌主要有两大功能。一是传达基本信息，包括店铺的名字和经营的类型；二是在店面结构中形成一个重要的设计元素。同样，店铺的标识也很重要，使用上根据位置、设计和数量不同，可以丰富店面的设计和街道的形象。（见图1.9）

关于招牌和标识的设计，有如下设计建议：

·招牌的尺度应该与店铺所在建筑的其他元素相互协调

·招牌或者突出的标识或广告牌不应遮挡建筑的开窗和细部

·招牌最好避免大面积使用丙烯酸（亚克力）等反光材料

·箱式招牌（一般采用塑料或金属材料）在现代店面设计中不常见

·标识可以出现在大面积的玻璃上，可以采用喷涂或者蚀刻的方法加在窗户内侧。但是要避免玻璃上出现过多标识，这样会让橱窗显得杂乱

·突出或悬垂标识必须简洁，只传递必要的信息

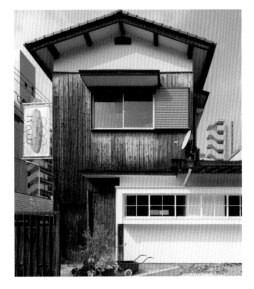

SHOP SIGNS

1.8 1.9

1.2 Branding Identity

Baked food is always a popular topic and you can find lovers of baked goods all over the world. Bakeries are fairly popular in the Western world and in recent years they spring out in Asian countries as well. When you present food in a delicious manner, you can almost expect business to come rushing through the door.

A bakery's brand is much more than a logo or slogan. It's the sum of everything the customer connects with – from the products being sold and how the bakery makes customers feel when they walk through the door.

Consumers often have compelling, favourable associations with strong brands and will often pay more or go out of their way for a strong brand. Brands are associated with quality, and quality commands a premium. Branding can give you a distinct advantage with specialty markets, can fortify a premium pricing strategy and can set your bakery apart in the minds and hearts of consumers.

In today's competitive bakery environment, differentiation is essential. Appealing to the consumer at these critical touch points can help consumers appreciate your bakery's brand and everything it stands for. (See figure 1.10~1.15)

Here are a few tips to help build a strong brand:

·Make a Connection with Customers

A bakery should aspire to connect with every customer on an emotional or experiential level. Each and every contact a customer has with your bakery matters. Pay extra attention to each aspect of the customer experience to add value to customers' impressions of your bakery and your brand. Remember, your brand can help build customer loyalty.

Some bakeries may not have branded products, but the experience of coming to your bakery might be part of your brand. Your brand may be as simple as being

· 招牌上的文字要清晰易读，风格要与建筑的其他部分相协调

· 简洁的文字和良好的对比效果会让标识更易读

1.2 品牌形象

烘焙食品是永远的流行话题，全世界都有烘焙食品爱好者。烘焙坊在西方非常流行，近年来在亚洲国家也开始大量出现。将烘焙食品摆放在精美的橱窗里营造出美味的样子，你就可以静候顾客盈门了。

烘焙坊的品牌形象不只是LOGO标识或者宣传标语那么简单。品牌形象是顾客见到的所有东西的总和，从待售的商品，到顾客进店时店铺给人的感觉。

顾客往往会青睐那些具有鲜明特色的品牌形象，他们会绕道去那样的店铺，消费也会更多。品牌形象由品质决定，品质意味着高档。品牌形象让你的店铺跟大众市场区别开来，占据优势，让你可以运用高价的经营策略，也让你的店铺在顾客心中变得与众不同。

在当今竞争激烈的烘焙坊市场环境下，与众不同是非常重要的一点。在上述重要方面让你的店铺对顾客形成吸引力，这样能够让他们欣赏你的品牌及其代表的品质。（见图1.10、图1.15）

以下是有助于树立品牌形象的一些小建议：

· 与顾客建立联系

烘焙坊应该尽量在感情和体验上与每位顾客建立联系。顾客与你的店铺产生的每次接触都非常重要。顾客体验的方方面面都需要注意，尽

Chapter 1
Shopfront and Branding Design

第一章
烘焙坊店面与品牌形象设计

1.10

1.11

known as the friendliest folks at the old-fashioned, delicious neighbourhood bakery.

Have clear expectations to welcome customers as they come through the front door. Offer assistance in finding the perfect baked treat, be positive and friendly.

·Be Part of the Community

Make sure you're personally connected to the community. Gain a better understanding of your community to help you to better serve your customers. Be visible by serving on local boards or having an active role in local events or committees. Work with a group that you believe in and has affinity with your bakery's goals.

Get credit for your involvement in the community. Promotions, sponsorships, programmes and particularly, your people can all significantly enhance your brand.

·Create a Niche

Finding a niche for your bakery means becoming known for offering just one or two baked goods or a signature item. If you don't know what item people like most from your bakery, ask them, and use that information to market a signature item. Differentiating your products by using certain ingredients, such as organic or fresh-ground flour is another way to build a niche. If you sell gluten-free, vegan, dairy-free or organic items, use that to create specialised niches.

·Customer Service

Becoming known as the bakery that welcomes every customer warmly while taking a second to chat with him helps you build a brand that focuses on customer service. Put your best people at the front counter, and train them to smile and welcome each person who walks into the bakery. If you offer free samples, teach your staff to offer a taste first, rather than wait for a customer to ask to try a sample.

Branding is of course, a very personalised thing. Specific ideas unique to your bakery and product mix must be explored. Understanding the importance of brand-

量提升顾客对店铺和品牌形象的印象。要记住，品牌形象能帮你建立顾客忠诚度。

有些烘焙坊可能没有品牌的产品，但是每次进店消费的体验本身就能形成你的店铺的品牌形象。顾客对你的烘焙坊的形象定位可能非常简单，比如说，附近街区内最友好的一家老式烘焙坊。

从顾客踏入店内的那一刻就要有充分的迎客准备。帮助顾客找到最完美的烘焙食品，态度要积极主动，亲切友好。

·成为社区的一部分

首先要确保你个人融入当地社区。要对你所在的社区有更好的理解，这样才能让你更好地为顾客服务。社区的活动要积极参与。要有固定接触的群体，让他们成为你的烘焙坊的老主顾。

参与社区生活能让你得到社区居民的认可。宣传、赞助、搞活动，包括你的固定顾客群体，都能帮你建立品牌形象。

·形成特色

如果你的店里有一两样特色产品，你的烘焙坊可能因此而为人所知。如果不知道你店内哪种产品最受欢迎，可以询问你的顾客，然后推出你的店铺特色产品。使用某些特别的原材料，比如有机面粉或者新鲜面粉，也能让你的店形成特色。如果你出售无谷胶、无牛奶添加的素食产品或有机产品，也可以以此作为特色。

·顾客服务

让你的烘焙坊以热情迎客而为人所知，花一点时间跟顾客聊几句。这样，能让你的店铺形成"关

1.12

1.13

1.14

1.15

ing, and doing everything in your power to enhance your branding experiences with customers is a key opportunity to grow your business.

注顾客服务"的品牌形象。最好的服务员要放在正面的柜台，让他们对进店的每位顾客微笑表示欢迎。如果有免费试吃，要让服务员主动提供，而不要等顾客去询问是否可以试吃。

当然，品牌形象是一种非常个性化的东西。一定要探索属于你的烘焙坊及其销售产品的独一无二的理念。要理解品牌的重要性，尽全力提升顾客的品牌体验，这是让你的烘焙坊发展壮大的关键所在。

Figure 1.1 The shopfront of Boulanger Kaiti, a bakery in Fukuoka, exudes simply elegance, a real eye-catcher in the area
图 1.1 日本福冈卡提面包店店面，简单但不失优雅，格外引人注目
Figure 1.2 Elements constitute the shopfront and the well proportioned scale
图 1.2 构成店面的要素
Figure 1.3 BREAD, ESPRESSO& by Aki Hamada Architects + Kentaro Fujimoto
图 1.3 Aki Hamada Architects + Kentaro Fujimoto 事务所设计的 BREAD, ESPRESSO& 面包店
Figure 1.4 Bakery MAISQUEPAN by NAN Architects
图 1.4 MAISQUEPAN 烘焙坊由 NAN Architects 事务所设计
Figure 1.5, figure 1.6 Style Bakery by SNARK
图 1.5，图 1.6：SNARK 事务所设计的风尚烘焙坊
Figure 1.7 The White Gift Box – Aimé Patisserie by LUKSTUDIO
图 1.7 Aimé 烘焙坊由 LUKSTUDIO 事务所打造
Figure 1.8 Boulanger Kaiti by MOVEDESIGN
图 1.8 MOVEDESIGN 事务所打造的 Boulanger Kaiti 烘焙坊
Figure 1.9 A well proportioned fascia
图 1.9 比例适中的招牌
Figure 1.10 to 1.15 Incorporating brand identity to every aspect related with the bakery to promote its image
图 1.10~ 图 1.15 将品牌形象融入烘焙店每个方面以推广自己的形象

A gift box along the street

矗立在路边的礼品盒

The unwrapping experience of the Aimé gift box is translated into the physical store. The idea of layering appears when the designers lift one semicircular translucent paper after another to discover the colourful macarons within. This opening sequence gives form to the overhead storefront design, while the window display made of four translucent layers attracts passers-by to explore inside the store.

Title/ 名称 : The White Gift Box – Aimé Patisserie
Designer/ 设计 : LUKSTUDIO
Photography/ 摄影 : Peter Dixie for LOTAN Architectural

设计概念来自拆开 Aimé Patisserie 包装盒的独特体验：捧着精美的盒子，一层层地打开透亮的半圆形包装纸，这揭示的过程让色彩缤纷的马卡龙更加吸引。设计师把这种乐趣呈现在店铺的设计中：店铺招牌的平面设计和橱窗的立体屏风设计，分别用这四层半圆的图案去吸引路上行人，带着好奇心进店内体验探索。

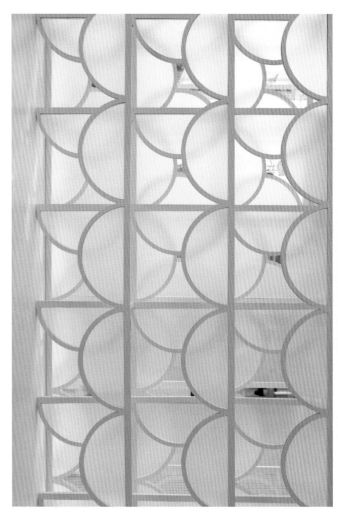

A bakery shop hidden in the residential houses

隐藏在居民区内的独特店面

The designers started to plan not to appeal the façade as a shop renewing the whole image of this building, but they regarded to match with the surroundings as important and let the shop fit this residential district, arranged façade design to integrate a new section into the old building, and furthermore, configured it so that people can recognise there is a bakery here.

The designers replaced only one surface of wall on south side with the shop section with all white appearance not to be imbalanced with the existing characteristics. Those wide lining windows attract interest of customers from the outside, from the inside, let the sun light moderately in and limit the view to outside so that customers can shop comfortably.

设计师不想改变整个建筑的面貌，不想让其看上去就很像一家面包店，而是想使其跟周围环境更加相容，跟整个住宅区的环境更协调。设计为原建筑增加了一个部分，同时也能让人们看出来这里有一家面包店。

设计师只改变了店铺部分南侧的外墙，改成全白，不会跟其他部分的特征相左。大面积的开窗能吸引注意力，包括外面的行人和店内的顾客，让阳光能温和地射入室内，同时也限制了向外观望的视野，让顾客能更舒适地选购。

Title/ 名称 : Boulanger Kaiti
Designer/ 设计 : MOVEDESIGN (Designer: Mikio Sakamoto)
Photography/ 摄影 : Yousuke Harigane (Techni Staff)

A modern and eye-striking bakery along the street

街对面的一家现代、引人注目的烘焙店面

Title/ 名称：Bakery MAISQUEPAN
Designer/ 设计：NAN Architects
Photography/ 摄影：Iván Casal Nieto
(www.ivancasalnieto.com)

"Our idea was to create a façade that was very striking, where the operation of the entire shop to be seen from the street – a very open place where the outer and inner communication was very fluid, made the stay in the premises in a pleasant experience and the street people feel encouraged to enter," the designers said.

A steel façade marked where interruptions emphasise the aspects they want to highlight. In this case an exhibition of wines, a bar where people look to the street and inside the counter, where you see people doing their own jobs of this type of establishment. Thus the door shown in the closed part searching with this contradiction highlights the other elements.

"我们的概念是打造一个富有视觉冲击感的店面，让街上的行人能看到整个店内的运营，实现室内外的流畅沟通，既为店内的人提供愉快的体验，又能吸引街上的人进入店内。"设计师如此解释设计。

不锈钢板立面突出了店面的设计，从街道上，人们能看到红酒展示、窗边吧台以及坐在店内各行其是的顾客和店员。封闭的店门与玻璃空间形成了对比，凸显了金属的质感。

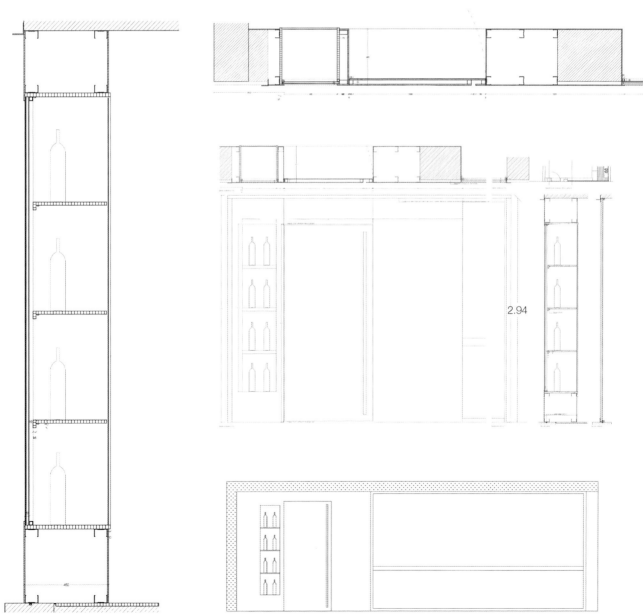

A unique design in a quiet alley

小巷弄里的独特设计

It is located on the ground floor facing an alley of apartment buildings. Since the street width is about only three metres, it gives greater impression as a façade when viewed from the side than from the front. To utilise this aspect, the façade was designed with dual characters, meaning it looks different when approaching from one end of the alley and from the other end, so that the shop could be an eye-catcher on the alley and characterise the alley's image. Triangularly arranged name boards, exterior walls, and planters were made of different materials, red cedar and cement board, for two sides. As a result, this brought unique characteristics to the façade when viewed from the front.

面包咖啡店坐落在一栋位于小巷的公寓楼的一楼。因为街道的宽度仅有 3 米宽，所以店面从侧面看去要比从正面看去更显眼。为了利用这一点，店面采用了双重设计，从小巷的两个方向分别会看到不同的店面招牌，这既让店铺变得更引人注目，又突出了小巷的特色。三角形招牌、外墙和花池的两面分别采用了红雪松和水泥板两种材料。从正面看去，小店显得别出心裁。

Title/名称：BREAD, ESPRESSO&
Designer/设计：Aki Hamada Architects + Kentaro Fujimoto

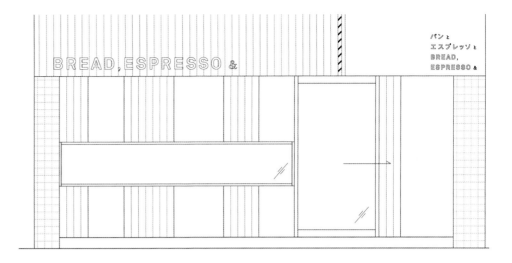

A distinctive renovation store at the square

广场前别致的翻新店铺

At the square, protected by its angel, the façade was discovered: it is the chocolate tile moulds used in chocolate making. Behind it, there is a vertical space that was born out of the effort to connect the five levels of the building. To achieve this, the designers rely on the staircase and its visual omnipresence as well as on the different passing gaps that link vertically the three lower levels.

店铺位于一个广场上,外立面的设计借鉴了制作巧克力要用到的模具。后面是个纵向空间,设计宗旨是让建筑的五个楼层连通。为此,设计师利用了楼梯。尤其是下部的三层,用楼梯实现了巧妙的视觉纵向关联。

Title/ 名称:FERRER XOCOLATA
Designer/ 设计:Arnau Vergés Tejero / arnau estudi d'arquitectura
Photography/ 摄影:Marc Torra_fragments.cat

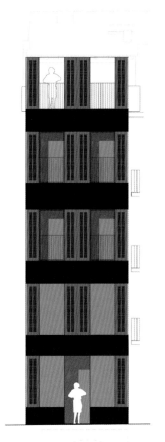
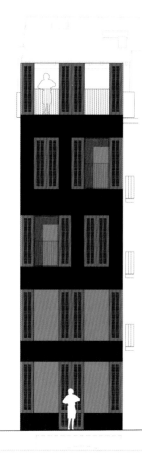
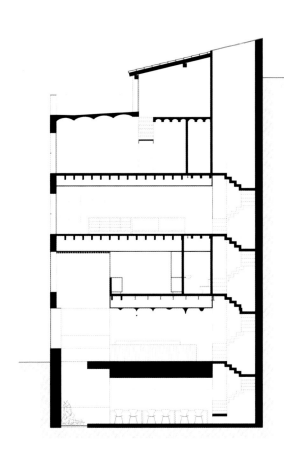
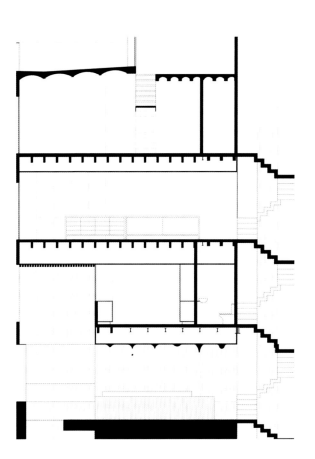

A shopfront full of visibility and modernity

良好的可见性，充满现代气息

The initial concept of the project was to reinstate the status of a buffet and patisserie with 25 years of market, while providing it with a new identity. La Galette's project makes an allusion to affection, singularity and refinement, seeking to portray the identity of old European patisseries with a modern touch.

Materials used are lettering in galvanised sheet, screen galvanised fence over pink-painted wall and ubatuba green granite on the pillars.

The expositor on the window shop, designed by the architect as two linear pieces on a sequence of lathed supports, also reveal the identity of La Galette, bringing an irreverent form of presentation. A handmade scissor window secures the window shop.

Title/名称：La Galette
Designer/设计：Architect David Guerra
Photography/摄影：Jomar Braganca

这家糕点店已有25年历史，设计宗旨是凸显其历史性，同时赋予其全新的品牌形象。设计着重表现了店铺的独特性、感染力、精致度，在老式欧洲糕点店的传统形象上增加了一丝现代的气息。

所用材料包括：镀锌钢板；粉色墙面上的镀锌栅栏；壁柱上的墨绿色花岗岩。

店面橱窗上的两个纵向板条是建筑师设计的，对店面形象的塑造也非常重要。手工打造的"剪刀窗"确保了橱窗的安全。

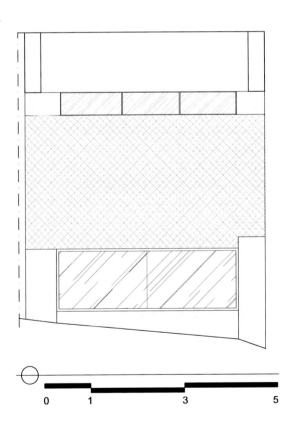
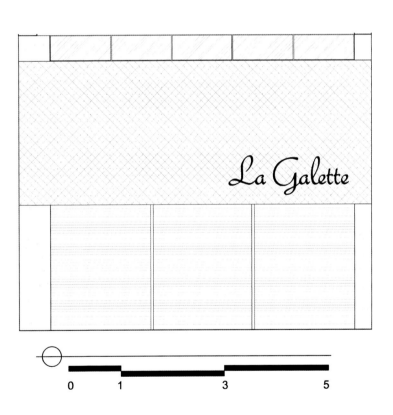

Preserving the historic flavour
保持历史风貌

The building is located in the historic centre of the town of Alcala de Henares. The bakery is on a ground floor of an interesting historic building. As soon as the designers demolished the internal walls and cleaned up the façade, they realised that they didn't have to invent that much. With their history, the existing brick walls spoke by themselves giving the space a huge personality.

这家烘焙坊位于西班牙阿尔卡拉城历史悠久的市中心区。店铺占据一栋古老建筑的一楼空间。设计师拆除了室内墙壁，对外立面进行了清理，之后发现，其实需要设计的并不多，保持原貌即可！砖墙散发出浓厚的历史气息，本身就能赋予店铺独特的个性。

Title/名称：Patisserie pan y pasteles
Designer/设计：ideoarquitectura
Photography/摄影：Miguel de Guzmán

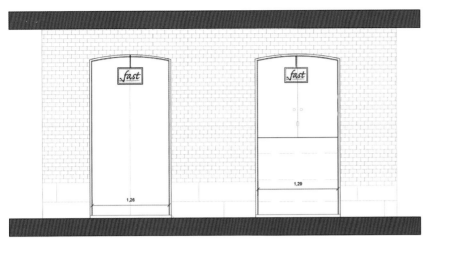
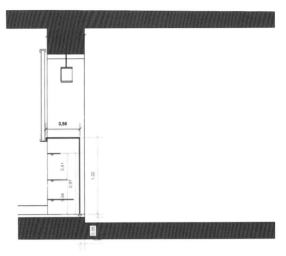
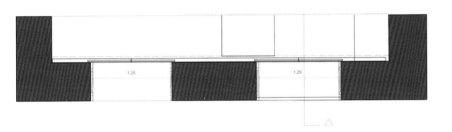
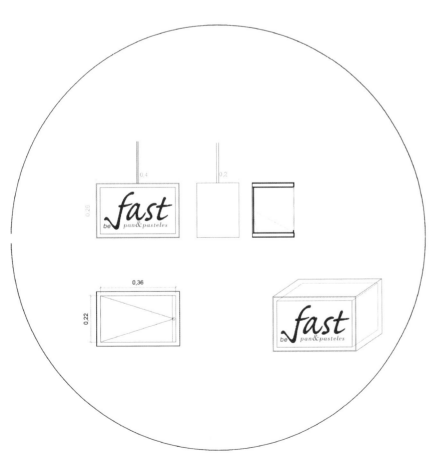

Preserving the endearing aspects of Germany's confectionery culture in a district of cross-section classes

在多阶层的街区内保留德国甜点文化的可爱之处

A district with history as the new home of the movement. Munich's Westend district houses a cross-section of social classes, all of whom leave their mark. Migrant workers, intelligentsia, students and tradespeople, coexisting in mutual understanding and respect.

The storefront highlights simplicity, with glass as the main glass, which gives the passerby a clear view inside. The name in pink and white is based on the branding colour, seeming like the sweet flavour emerging inside.

位于慕尼黑的 Westend 街区是一个各阶层社会人群的汇聚地，移民工人、知识分子、学生、商人共同生活在这里，他们互相理解和尊重。每个群体都在这里留下深深的印迹。

店面设计以"简约"为宗旨，通透的玻璃作为主要材质，邀请路边的行人走进来。店铺名字采用粉色和白色，既突出品牌形象，又似乎散发着店内甜品的芳香。

Title/ 名称：Das Neue Kubitscheck
Designer/ 设计：Designliga
Photography/ 摄影：Designliga

Corporate Design

Multifaceted and dynamic. Free from rigid frameworks or visual museum pieces. The corporate design needs to be vibrant, in a state of permanent flux and change, adapting to developments instead of compelling alignment. The focus is the statement 'Fuck the Backmischung', as the journal, menu and mouthpiece of the movement.

企业形象设计

企业形象的设计是多方面且充满活力的，超越了条条框框和俗套的视觉设计。企业设计必须生机勃勃，处在永恒的变化之中，随时应对发展需求，而不是一成不变。设计的重点是"颠覆烘焙"，这句口号被广泛应用在杂志、菜单和活动宣传中。

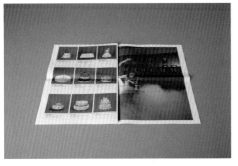  
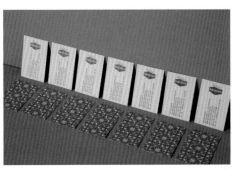  

Visual branding inspired by the art of Lithuania and Belarus

源于立陶宛和白俄罗斯的视觉形象设计

Karaway is a brand with a modern sensibility that takes inspiration from Easter European traditions. The name Karaway is derived from the Slavic greeting Karavi, a custom of welcoming guests with a loaf of homemade bread. The illuminated letter 'K', red, and the brand pattern were all inspired by the art of Lithuania and Belarus. To complete the visual brand, Cristian Barnett was commissioned to photograph in the Karaway bakery, perfectly capturing the baking process and the breads.

ico collaborated with architects 31/44 to develop the first Karaway shop, where the use of natural materials continues in-store, from brown-paper bags to linen staff aprons and handwritten price tags. The open-plan shop creates the feel of a market, using wooden food crates and wicker baskets to display goods.

Karaway 是一个植根于东欧传统，同时又具有现代气息的品牌店铺。店铺的名字来源于斯拉夫人的招待礼仪"Karavi"——斯拉夫人待客时会奉上一条自制的面包。店面上一个大大的红色"K"字母，以及与其品牌形象相呼应的图案，设计灵感都源于立陶宛和白俄罗斯艺术。为完善品牌视觉形象设计，设计师克里斯蒂安·巴奈特还特意来到店内摄影，亲自见证了面包烘焙的过程。

ico 设计公司联手 31/44 建筑事务所，共同打造了 Karaway 的首家店铺。设计使用了天然材料，从棕色纸袋到店员的亚麻围裙以及手写的价签。店内采用开放式布局，营造出一种集市的感觉，采用木质板条箱和柳条篮来盛放和展示食品。

Title/ 名称：Karaway
Brand, design & art direction/ 品牌设计与艺术总监：ico Design
Interior designer/ 室内设计：31/44
Photography/ 摄影：Cristian Barnett
Location/ 地点：London, UK/ 英国，伦敦

Communicating bakery's philosophy through identity design

用品牌形象设计传达烘焙坊的生活哲学

Abarrotes Delirio balances street life and good food through its authentic and functional offer, conveying a lifestyle which is impulsed by a new gastronomical street culture and gourmet meals, but in a new pantry-ready format. The objective was to communicate Abarrotes Delirio's philosophy, based on practicality, simplicity and on the organic provenance of their ingredients, through its identity and interior design, which in turn was developed by Habitación 116.

Abarrotes Delirio 烘焙坊将美食与街头生活相结合，以一种全新的方式表现了一种新型的街头美食文化。设计目标是传达出 Abarrotes Delirio 烘焙坊的生活理念，简单，实用，健康的原材料。品牌形象设计和室内设计由 Habitación 116 设计公司操刀。

Title/ 名称：Abarrotes Delirio
Design agency/ 设计：Savvy Studio
Creative director/ 创意总监：Rafael Prieto & Raul Salazar
Designer/ 设计师：Eduardo Hernandez
Photography/ 摄影：Alejandro Cartagena
Client/ 客户：Monica Patiño
Location/ 地点：México/ 墨西哥

Cocinamos con aceite de oliva de Ensenada, sal de mar de Cuyutlán y especias de todo el mundo. Todos nuestros productos son hechos con dedicación : usando sólo la mejor materia prima, con especial interés en su proveniencia.

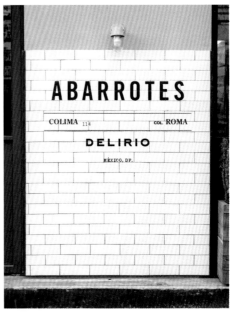

Website design to connect delicious baked goods and the story behind them

用网站设计将美味的烘焙食品及其背后的故事表现出来……

Martin Auer is known for thinking outside the box – or rather the baking tray – and it was important to highlight that the whole brand is about more than bread: it is a lifestyle, an appeal and a masterly craft. The website is creating an emotional connection between the delicious baked goods and the story behind their creation: large, impressive photos bring the special devotion to baking to life alongside humourous illustrations. You experience a sympathetic, modern and innovative brand and the person behind it – Martin Auer, who we honestly believe wants to give the bread its soul back.

Title/ 名称：Martin Auer
Design agency/ 设计公司：moodley brand identity, grafisches Büro
Designer/ 设计：Nicole Lugitsch
Photography/ 摄影：Michael Königshofer
Client/ 客户：Martin Auer GmbH
Location/ 地点：Austria/ 澳大利亚

马丁·奥尔品牌烘焙坊以其"烘焙之外的思考"而闻名，因此，这个项目的设计要体现出这样的理念：马丁·奥尔品牌，不只是烘焙。它是一种生活方式，一种极具感染力的艺术。网站的设计旨在将美味的烘焙食品与其背后有关烘焙过程的故事联系起来。巨幅照片极具视觉冲击力，加上幽默的插图，充分表现出人们对烘焙食品的热爱。在这里，你会体验到一个现代的、创意的品牌，进而去了解这个品牌背后的人——马丁·奥尔，一个致力于"让灵魂回归面包"的人。

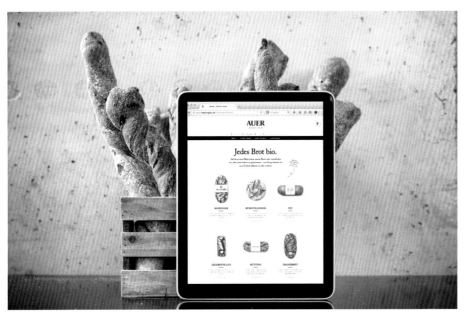

Corporate identity to communicate the best homemade touch of the bakery

用品牌形象设计表现自制烘焙食品的独特韵味

Susy's Bakery is a premium quality bakery and food retail space founded and established by Azucena Romero Camarena since 1976 in Guadalajara, Mexico. The corporate identity is directly derived from the profile of the company: a small business which bakes signature gourmet cookies, cakes, cupcakes, pies, and choux, priding itself of having the best homemade touch of the region.

Susy's Bakery's packaging is quite simple and very easy to apply; they use parchment paper to wrap the different products, which is printed with a pattern of pictograms specially designed for the brand. Circular stickers are also printed with pictograms to stick on laminated packaging; finally, when delivering the client their purchase, recycled paper bags will be used, printed with different designs, each made for small bags and for larger bags.

苏西烘焙坊是一家高档烘焙食品零售店，位于墨西哥西部城市瓜达拉哈拉，由阿祖切纳·罗梅罗·卡马雷纳创立于 1976 年。品牌形象直接来源于公司自身的定位：一家制作特色曲奇、蛋糕、馅饼和甜品的小店，为能制作该地区最好的自制烘焙食品而自豪。

苏西烘焙坊的食品包装非常简单，容易操作。各种食品的包装都采用羊皮纸，上面印有专为这家店铺设计的品牌形象图形。贴在包装纸上的圆形贴纸上也有品牌形象图形。最后，将食品交给买家的时候，店铺会使用回收的纸袋，上面印有不同的图案，分大小两种。

Title/ 名称：Susy's Bakery
Design agency/ 设计公司：Memo & Moi Brand Consultants
Creative director/ 创意总监：Guillermo Castellanos & Moisés Guillén
Designer/ 设计：Moisés Guillén & Guillermo Castellanos
Photography/ 摄影：Jani Rangel Lucero
Client/ 客户：Azucena Romero Camarena
Location/ 地点：Guadalajara / México/ 墨西哥，瓜达拉哈拉

Identity design inspired by Buenos Aires' coat of arms

灵感来自布宜诺斯艾利斯盾徽的品牌形象设计

Anagrama designed Violeta's new identity with the mission to communicate quality and sophistication without losing approachability.

The proposal draws inspiration from Buenos Aires' coat of arms: an oval with two ships sailing over Rio de la Plata and a dove flying above them. They multiplied the dove by three in celebration of Violeta's three decades of fantastic bread and kept the lines depicting the Rio de la Plata and used them not only in the icon, but all over the brand in four different patterned textures. Violet was chosen to match the naming and to keep the brand fresh and feminine. The copper foil is not only a nod towards Violeta's excellence, but it also refers to the toasty colour of bread right after it leaves the oven.

Title/ 名称：Violeta
Design agency/ 设计公司：Anagrama
Client/ 客户：T4TURBAN
Location/ 地点：Argentina/ 阿根廷

Anagrama 设计公司为维奥莱塔设计了新的品牌形象，设计宗旨是传递出该品牌对高品质的不懈追求，同时又不失亲民性。

设计的灵感来自布宜诺斯艾利斯盾徽的纹章，一个椭圆形的标识，上面有两艘帆船，在拉普拉塔河上航行，还有一只鸽子，在上面飞过。设计采用了三只鸽子的形象，寓意是维奥莱塔烘焙坊有 30 年制作精品烘焙食品的历史。其余线条仍保持拉普拉塔河的形象，不仅用于品牌标识形象，而且以四种不同图案的形式体现在品牌的方方面面。店铺的名字选用了紫罗兰色，清新淡雅，充满女性的柔美。铜箔的使用不仅体现了维奥莱塔品牌的超高品质，而且正好也是面包刚出烤箱时的颜色。

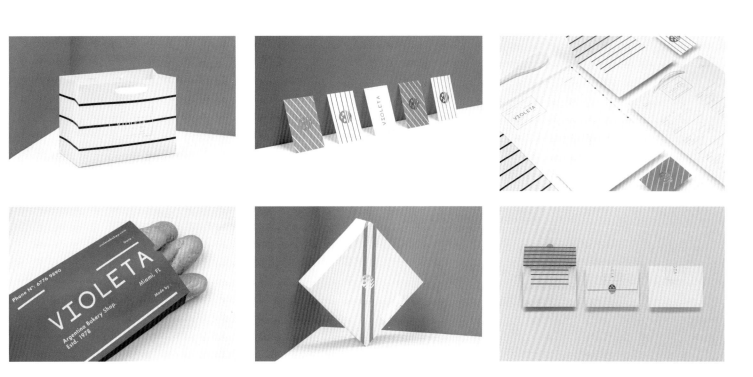

Identity based on unique illustrations and brand character

以独特的插画和品牌特色为出发点的品牌形象设计

The smell changes life. Backbone Branding developed a brand character – Louis Charden, the founder of the bakery, and created a brand story. The identity was based on unique illustrations, unfolding different life events of the brand character – his first love, childish amusements, his first steps in the confectionary industry.

味道改变生活。"品牌支柱"设计公司(Backbone Branding)为路易·查登烘焙坊打造了其品牌创始人路易·查登的卡通形象,并用其讲述了该品牌背后的故事。这个品牌形象以独特的插画为出发点,以插画的形式展现了路易·查登不同的生活片段:初恋、孩子气的消遣、初次踏入糕点制作领域的历程,等等。

Title/ 名称 : Louis Charden café and bakery
Design agency/ 设计公司 : Backbone Branding
Art director/ 艺术总监 : Stepan Azaryan
Designer/ 设计 : Karen Gevorgyan
Illustrator/ 插图 : Anahit Margaryan
Client/ 客户 : Karen Margaryan
Photography/ 摄影 : Backbone Branding
Location/ 地点 : Armenia/ 亚美尼亚

Chapter 2 Space Layout & Interior Design
第二章 空间布局与设计

2.1 Space Layout
2.1 空间布局
2.2 Interior Design
2.2 室内设计

Case Study
案例解析

Bulka Cafe and Bakery
布尔卡咖啡烘焙坊

LOISIR
洛伊斯烘焙坊

hey ju
"嘿!宇"烘焙工坊

The White Gift Box - Aimé Patisserie
Aimé 烘焙坊

Bakery Marie Antoinette
玛丽皇后烘焙坊

Parken Bakery
帕肯烘焙坊

The Cake
蛋糕店

Royal Bakery
皇家烘焙坊

Patisserie Pan y Pasteles
西班牙马德里 Pan y Pasteles 甜品店

Vyta Santa Margherita Bakery
圣玛格丽塔烘焙坊

Zhengzhou Industrial Style Bakery
郑州工业风烘焙店

Kao Nekad Restaurant & Bakery
旧时光烘焙餐厅

Bakery MAISQUEPAN
梅斯科潘烘焙坊

Bartkowscy Bakery
巴特科沃斯基烘焙坊

BREAD, ESPRESSO&
BREAD, ESPRESSO& 面包咖啡店

BINARIO 11
拜纳里奥 11 号

DESIGNER'S VISION - Colour and forms serve as 'input' to attract customers walking in
设计构想——以色彩与形态来吸引顾客走进来

Chapter 2
Space Layout & Interior Design

第二章
空间布局与设计

2.1

2.2

Many people might think customers will visit a bakery simply because of the quality of its goods, but the space of a bakery and interior design are equally important in attracting new and old customers and adding to the overall shopping experience. The space planning, use of colour, materials, lighting, and decorative features have an impact on customer experience.

Making a first impression is crucial in business, and the only way to make a good first impression is not through the food, but what the customers see as soon as they walk through the door. (See figure 2.1, figure 2.2)

Interior design is an integral part of the overall success of a bakery and must not be overlooked. The followings are what the client as well as the designer must consider:

·What kind of shopping experience do you want customers to have when they walk into the bakery?

·Does the interior design and layout convey that experience?

·Does the bakery provide an easy flow for customer traffic and service?

·Are the kitchen and workstations ergonomically designed to bring out the best in staff?

2.1 Space Layout

Starting a new business is always a daunting experience and the shop layout will understandably be the last thing on your mind. It is very important to carefully think through the layout of a bakery shop. Here are some ideas to guide you.

Keep customer service area spacious and inviting

Customers want to feel relaxed about the place in which they have chosen to spend their money. This should happen for them as soon as they walk through

很多人可能认为，人们会光临一家烘焙坊是因为店里商品的质量，但是烘焙坊的室内空间和设计也同样重要，能够吸引新老顾客，提升整体购物体验。空间规划、色彩运用、材料、照明以及装饰元素都会对顾客的购物体验产生影响。

对于商业经营场所来说，留下良好的第一印象十分重要，而烘焙坊留下第一印象的方式却不是食物，而是顾客进门后首先看到的店内环境。（见图2.1、图2.2）

室内设计是一家烘焙坊整体成功的关键一环，一定不能忽视。以下几点是店主和设计师都要考虑的：

·你想让顾客走入店内后拥有什么样的购物体验？

·室内布局和设计是否支持你对购物体验的预期？

·烘焙坊室内的交通和服务动线是否流畅？

·厨房和工作台是否符合人体工学设计，让工作人员能以最好的状态投入工作？

2.1 空间布局

要新开一家店总是一个令人犹豫踟蹰的过程，最后才会想到店内空间布局，这也可以理解。但是，烘焙坊的室内布局绝对需要精心推敲。下面提供一些布局的建议。

顾客服务空间要宽敞宜人

顾客希望在他们选择消费的店里感到放松。他们一走进店内就应该有这种感觉。这就意味着，

2.3   2.4

the doors. This means keeping the entrance roomy and free of clutter. This can be achieved by using Slat Wall Hooks to put the space on walls to good use while freeing up space for chairs, standing areas, etc. How many times have you heard people decide against visiting a shop due to crowd or lack of space? Quality baking will help boost sales but an uncluttered shop will win lots of return customers.

If there is enough space, offer a place where customers can put down their other bags of heavy groceries that they may have just purchased; this will make their browsing and buying time more enjoyable and comfortable.

If the bakery is located inside a building, keep the main doors open wide to offer as much of a view of the interior as possible – and also to let the aroma of freshly baked goods flow out of the store and attract attention.

Moreover, it requires much attention to locate the cashier – the situation of a long queue should be avoided. There are many modern payment methods that can be applied, for example, pay online. It is an effective way for both customers and the bakery. (See figure 2.3, figure 2.4)

Be focus on display area

Functionality is a huge necessity when working on the layout of a bakery shop. Customers will be drawn in by the baked goods. This means to make the goods the focal point of the shop. Design the layout so that goods are placed strategically around the door and window areas. All machinery and tools should take the back seat away from the customers' line of sight.

If the space allows, place counters and cases facing the door and storefront to display your baked goods to window shoppers and draw in customers. The next best layout puts the glass cases perpendicular to the door, so that the end of a case displays baked goods near the entrance.

入口处的空间要宽敞，没有障碍物。可以使用墙面挂钩来尽量利用空间，将地面空间解放出来，放置椅子或者作为站立空间。你是否无数次听人说起过，某家店太拥挤、空间太小，所以不想再去？美味的面包当然会有助于生意兴隆，但宽敞的室内空间会给你带来更多的回头客。

如果空间足够大的话，设置一个地方让顾客可以放包或者在别的地方刚采购的物品，这样会让他们在店内选购的过程更加轻松。

如果烘焙坊不是独栋结构而是在某建筑内，要保证大门足够开敞，从外面可以看见室内，也让新鲜面包的香气飘出店外，吸引更多注意。

此外，收银台位置的选择也要注意，尽量避免排长队。有很多现代的付款方式可以采纳，比如，网上支付，对顾客和店家都非常方便。(见图2.3、图2.4)

展示区是重点

对于烘焙坊的空间布局来说，功能性是最基本的要求。顾客会被烘焙食品吸引进店。这意味着食品应该是店内的主角。空间布局要让食品能在门口和窗户前进行展示。设备和工具应放置在后方，在顾客的视线范围以外。

如果空间允许的话，柜台和陈列柜可以面向大门和店面，让过路人都能看到美味的食物，吸引他们进来。还有一种布局也很好：将玻璃陈列柜布置为垂直于大门，这样，陈列柜尾端就在靠近入口的位置。

展示区的设计还要注意顾客交通动线的流畅，确保他们能够舒适愉悦地享受购物的过程。(见图2.5~图2.7)

Chapter 2
Space Layout & Interior Design

第二章
空间布局与设计

2.5

2.6

In display area, it is also very important to keep a fluent movement line for customers, letting them select goods in a comfortable and pleasing way. (See figure 2.5 to figure 2.7)

Reserve adequate space for staff

The staff space usually includes baking kitchen, storage room, fitting room, etc. Usually, (especially in the past) those spaces are out of the eyes of customers. However, as the dining habit changes, most bakeries decide to set up those spaces in front, especially for the baking kitchen to perceive the whole process as a performance to attract customers. As for the other spaces, they should be designated based on the entire layout of the bakery, but must remember to keep them adequate. (See figure 2.8)

2.2 Interior Design

Choose the right colour scheme

Bakeries are a popular place to have a treat and spend time with friends and family. For a bakery, the colours to decorate with are more important than you think. Colour plays a role in mood and appetite, making the right choice vital for running a successful bakery. Here are some rules to follow:

* Look at the logo and advertising materials

Matching the interior colour of a bakery with these items allows customers to associate them, which could bring in more business. For example, if the sign and business cards feature sherbet colours, like pale pink and yellow, the same interior associates the two, allowing potential and return customers to see your sign and want your treats.

* Watch out for the size of the bakery

Darker colours are not good for small spaces as they will make the space feel

确保员工有足够空间

员工空间通常包括：烘焙厨房、储藏间、更衣室等。一般来说（尤其是过去），这些空间是不在顾客的视线范围内的。然而，随着我们用餐习惯的改变，现在大多数的烘焙坊都会选择将这些空间搬到前台，尤其是烘焙厨房，能展示整个烘焙的过程，作为吸引顾客的一种手段。至于其他员工空间，可以根据烘焙坊整体的空间布局来布置，但要确保空间足够大。（见图2.8）

2.2 室内设计

选择适当的色彩

烘焙坊是日常请客或者与亲朋好友小聚的好地方。烘焙坊的装饰色彩是门学问，远比你想象的更复杂。色彩会影响情绪和食欲，因此，选择适当的色彩对于烘焙坊的成功经营来说尤为重要。以下是设计中要遵循的一些法则：

* 注意LOGO标识和宣传广告

烘焙坊的室内色彩要与店铺的LOGO标识和广告宣传保持一致，便于顾客进行自然的关联，这样能带来更多的生意。比如说，如果店铺的名字和卡片是甜美柔和的淡粉红色或淡黄色，室内设计也可以用这两种色彩，统一的色调能吸引更多潜在顾客和回头客。

* 注意烘焙坊的规模

较深的色彩不适合小面积的烘焙坊，因为深色会让空间显得更小。相反，浅色会让店内显得温暖、宽敞。如果是一家高档烘焙坊，可以考虑使用柔和的中性色。亮色能营造出空间的趣味性。在实际应用前建议使用颜色板来挑选哪

2.7

2.8

2.9

2.10

2.11

even smaller. Lighter colours on the other hand will add warmth and space to the shop. If the bakery is an upscale one, consider using soft and neutral colours. A fun atmosphere can be achieved with brighter hues and colours. A good tip is to use samples and swatches to determine what colours match the desired style and impression before actual use.

* Think about the style of the bakery

Two or three types of colours can be selected as the main tone based on the style of the interior. For example, the Nordic style usually features the colour of natural wood and brick. For the modern style, light colours often used, while for the traditional bakery, warm colours such as yellow, orange and red are mostly applied. (See figure 2.9 to figure 2.11)

种颜色最适合预期的风格和氛围。

* 考虑烘焙坊的风格

可以根据烘焙坊室内的风格选择二到三种色彩作为主色调。比如，北欧风格通常选用天然木材和砖石的色彩。而现代风格则常用浅色。暖色适合传统风格的烘焙坊，比如黄色、橘色和红色，是最常用的颜色。（见图2.9~图2.11）

对于不同的烘焙坊来说，色彩选择和配色可能有很大差异，以下是针对烘焙坊色彩设计的一些建议：

· 选择一种能够增进食欲的色彩。绿色、红色、棕色和橘色是几种能让人感到饥饿的色彩，因为这几种颜色使人联想起美味的食物——比如

Chapter 2
Space Layout & Interior Design

第二章
空间布局与设计

Colour selection and matching differ a lot in different bakeries, and here are some tips helpful to a bakery colour scheme:

·Choose a colour that increases appetite. Shades of green, red, brown and orange are colours that may make customers feel hungry, because they mimic the colours of appetizing foods such as chocolate, leading them to purchase more. Use these colours on the walls, seat cushions, display cases and wall art. In addition, bright orange, red and yellow stimulate the urge to eat, reports CNN. These colours can be added as accents or as the main colours.

·Use colours that promote a good mood. Cheerful customers are more likely to be comfortable in the bakery. This means they may linger longer and might spend more money, according to CNN. Yellow, orange and red are colours that invoke joy, energy and excitement, while blue and green are relaxing and calming.

· Stay away from dark colours. Deep blues, purples, or blacks may not be a good choice for a bakery. These colours aren't associated with an increase in appetite.

· Try white. White tends to open up small spaces, making it an ideal choice for a small bakery. It also gives a blank canvas for mixing up the decor with the seasons and holidays.

· Light-coloured walls, such as white, cream, pale yellow, pine or maple, create a spacious effect. A simple colour palette with one dominant colour and one accent colour, such as pale yellow with burnt orange, will help keep a small bakery from looking too cluttered.

· Test colour choices. Gather paint samples and fabric swatches, and display them in the bakery. This will help determine if the choices will work with the space and lighting. This also makes it easier to change things if you aren't happy with the look of the bakery, without going over budget and having to start from scratch.

巧克力，自然也就会诱导人买的更多。墙面、椅垫、陈列柜和饰品可以选用这些色彩。另外，根据CNN（美国有线电视新闻网）的报道，橙黄色、亮红色和嫩黄色能刺激人的食欲。可以加上这些色彩作为空间中突出的焦点色或者主色调。

· 使用能够激发良好情绪的颜色。据CNN的报道，轻松愉悦的氛围更容易让顾客在店内感到舒适。这就意味着，他们可能会待更长时间，花更多钱。黄色、橘色和红色能令人感到愉悦、充满活力、精神振奋，而蓝色和绿色则令人放松、冷静。

· 远离深色。深蓝色、深紫色或者黑色对于烘焙坊来说可能不是好的选择。这些颜色无法与增进食欲联系起来。

· 尝试白色。白色能让小空间显得更宽敞，因此，是小型烘焙坊的理想选择。白色也相当于一片空白的画布，在上面可以根据季节和节日进行自由的装饰布置。

· 浅色的墙面，比如白色、奶油色、浅黄色、松绿色或者淡棕色，能够营造出宽敞的感觉。简单的配色（一种主色加一种焦点色，比如浅黄色加焦橙色）能让空间较小的烘焙坊也不会显得凌乱。

· 色彩测试。拿到涂料样品和颜色布板，陈列在烘焙坊里，这样能够比较和测试某些选择是否跟室内空间和照明效果相协调。而且这样的话，如果你对效果不满意，要改也很方便，不用重新考虑预算再从头开始。

照明是关键

Lighting is key

The interior lighting is the most important design feature for any interior design project. It should be an integral part of the whole design, showcasing the best features of a bakery. Pay attention to where there is natural light at different times of the day before selecting and installing light fixtures.

Include different creative light fixtures to enhance diverse focal areas even if there is a small budget. Remember: the best lighting doesn't call attention to itself. The eye is drawn to the area being lit, rather than the light source.

*Keep in mind of the intensity

Is the lighting bright or is it dim? No one really over-lights their bakery. Problems are usually due to the east- or west-facing windows. If the sun shines directly through windows at any point during the day, some curtains or shades provide a quick and easy fix. Not only will customers be able to take their sunglasses off while sitting down, some light-weight curtains or shades will diffuse that sunlight and leave the bakery looking movie-set ready.

Don't forget about the production area when it comes to lighting. Sometimes these areas get overlooked in a bakery. (Walk into any art studio and observe the lighting design. You'll notice that artists typically work in well-lit and evenly lit spaces.)

*Create a pleasant environment

Creating an environment through good lighting is not only conducive to the output of employees, but also to that of customers' experience.

Bakery production is a visual medium. The production area is the studio where pastry chefs, bakers and decorators need lighting that provides the best environment for their work.

室内照明对于任何室内设计项目来说都是至关重要的一环。照明应该是整体设计中不可分割的一部分，突出展示店内的特色。要注意，一天之中哪些时段、哪些地方有自然采光，然后再选择和安装灯具。

即使预算有限，也要选择不同的创意照明方式来突出店内的重点区域。记住：最好的照明不是让人注意照明本身，目光是要吸引到被照亮的地方，而不是光源自身。

* 注意照明强度

照明效果要明亮还是昏暗？事实上，烘焙坊的照明可以说是多亮都不为过。常见的问题是东侧和西侧的开窗。如果白天日光直接从窗口照射进来，可以用窗帘或百叶窗来快速、方便地解决这一问题。这样，顾客进店坐下之后就可以摘掉太阳眼镜，而且窗帘或者百叶窗还能起到漫射日光的作用，让烘焙坊的室内形成电影中一般曼妙的光影效果。

照明设计不要忽略了烘焙坊的烘焙区。这些地方有时得不到重视。(走进任何一间艺术工作室，看看那里的照明，你会发现，艺术家都是在照明均匀而良好的空间中工作。)

* 打造令人愉悦的环境

利用良好的照明打造令人愉悦的环境，不仅有助于提高店内员工的工作效率，而且也能提升顾客的购物体验。

面包的烘焙是一种视觉媒介。烘焙区就相当于一个工作室，糕点师和烘焙师需要适当的照明为他们营造出最好的工作环境。

Chapter 2
Space Layout & Interior Design

第二章
空间布局与设计

2.12

2.13

Straining eyes in order to see detail work with a piping bag or taking Tylenol daily to relieve fluorescent lighting headaches while molding concha dough is not necessarily formulas for success, or a happy thing. Ideally, give them windows. If windows are not an option for production area, the daylight-balanced LED bulbs and fixtures will give a head start on making the production area a better lit environment for employees. (See figure 2.12, figure 2.13)

Choose alluring decorating items

Decorating the interior space of bakery is one of the ways – along with the bakery name, physical location and menu – that the choices will influence the way the customer perceives the bakery. For example, a bakery with a down-home, rustic style heavy on reclaimed barn wood will be perceived differently than a bakery with a European name and a velvet-and-glass aesthetic reminiscent of an upscale jewellery store. No matter what overall style being selected, borrow a cup of decorating inspiration from the items the bakery will sell.

*Fresh flowers

Place vases or simple glass mason jars full of fresh flowers throughout the bakery. Choose non-scented or mildly scented flowers so they don't interfere with customers' enjoyment of the food smells. Pick flower types that match the colour scheme. Not only are flowers an inexpensive form of decor that can be changed to reflect different seasons, but also they create a perception of freshness that customers may assume extends to the baked goods.

*Quote

Paint a quote related to baked goods, sweet goods, eating or food in general in large letters on one wall of the bakery. Alternately, you could stencil the quote along the top of the bakery walls, forming a thoughtful border, or have a selection of quotes printed on canvas, stretched over frames and hung throughout the bak-

员工眯起眼睛才能看清细节（比如裱花袋）？烘焙师傅每天得吃药来缓解制作糕点时荧光灯照明带来的头痛？成功的烘焙坊不一定要靠这些。最理想的方法是用开窗来解决。如果对于烘焙区来说开窗不可取，LED灯可以代替日光，安装在头顶，让烘焙区成为一个令员工感到舒适的照明环境。（见图2.12、图2.13）

选择有趣的装饰元素

除了烘焙坊的名字、地理位置、菜单等方面之外，烘焙坊的装饰也是能影响顾客感观的重要因素。比如说，一家烘焙坊用回收的旧木板营造出乡村风格，另一家取个比较欧化的名字，材料使用天鹅绒和玻璃，营造出一种高端珠宝店的感觉，那么这两家店就会给人完全不同的体验。不管选择哪种整体风格，在装饰方面，都可以从烘焙坊中出售的食品上寻求灵感。

* 鲜花

可以在烘焙坊各处都摆上花瓶或者简单的玻璃罐，里面插满鲜花。选择无香味或者香味比较淡的花卉，不要让花香影响到顾客对食品香味的感受。花卉品种的选择要与店内色彩相协调。花卉不仅是一种廉价的装饰元素，可以随着季节来更换，而且，还可以营造出一种新鲜与活力的感觉，让顾客很自然地感觉这种新鲜和活力，也延续到店内的烘焙食品上。

* 引用文字

可以引用与烘焙、甜品、饮食有关的文字，全大写喷绘在墙上。或者，你也可以把文字作为墙壁顶端的边线，或者把文字印在帆布上，再用框架绷开，挂在店内。引用的文字可以是：

2.14

2.15

2.16

ery. A few quotes such as 'Tell me what you eat, and I will tell you who you are' or 'All happiness depends on a leisurely breakfast', can be used.

*Portrait gallery

Bedeck the bakery walls with framed photo portraits of all the different foods and beverages being served. Hire someone with food photography experience to capture the mouth-watering essence of each item, from a steaming mug of black coffee to a fresh, colourful fruit tart to a blueberry crumble muffin, its top glittering with sugar. Showcase each item against a plain background. Frame the photos in identical frames and hang them art-gallery-style – in a single, eye-level row – around the bakery.

*Whimsical mural

Paint a whimsical mural on one wall of the bakery. This decorating idea is particularly appropriate for bakeries with lighthearted, brightly coloured decor. Hire a local artist to deck the walls with smiling cupcakes, muffins and bagels frolicking in hot chocolate fountains. Alternately, if the bakery has a more genteel, European feel to it, commission a mural that depicts bakery goods in the style of a European master painter. (See figure 2.14 to figure 2.16)

There are numerous items that can be selected to decorate the bakery and the aim is to make it more appealing to customers.

告诉我你吃的什么东西，我就知道你是什么样的人；所有的幸福取决于一顿悠闲的早餐。

* 摄影装饰

烘焙坊的墙面可以用画框来装饰，摄影的图片可以是店内的各种食品和饮料。可以请一个有食品摄影经验的人，拍摄出每样美食令人垂涎欲滴的瞬间：一杯冒着热气的清咖啡、一个色彩缤纷的新鲜水果蛋挞、一块撒了晶莹糖粒的蓝莓松糕……拍摄物的背景要简单。照片装在同样的画框里，像美术馆那样挂在周围的墙上：单排，与视线齐平。

* 壁画

可以在烘焙坊的一面墙上画上吸引眼球的壁画。这样的装饰方法尤其适合装饰风格轻松、色彩明快的店铺。可以请一位当地的画家，在墙上画出纸杯蛋糕、松糕、百吉饼在热巧克力喷泉里嬉戏的画面。或者，如果你的烘焙坊是比较文雅的欧式风格，你也可以订制一幅欧洲绘画大师风格的烘焙食品画。（见图 2.14~ 图 2.16）

烘焙坊的装饰有无数的元素可供选择，总之，目标就是让你的店铺对顾客更有吸引力。

Figure 2.1, figure 2.2 Both the two bakeries create a pleasant atmosphere to attract customers in
图 2.1、图 2.2 这两家烘焙坊店面都营造了宜人的环境，吸引着顾客的到来
Figure 2.3, figure 2.4 The clean and comfortable customer area enlightens the entire bakery
图 2.3、图 2.4 整洁而舒适的顾客区提升了整个烘焙坊的品质
Figure 2.5 to 2.7 The display area serves as a central place in the bakery
图 2.5—2.7 陈列区在烘焙坊中占据着重要的作用
Figure 2.8 The open kitchen provides a stage for the staff and lets customers enjoy the making process of the delicious baked products
图 2.8 开放式的厨房为员工提供了发挥的舞台，同时也能让顾客享受美味食品的诞生过程
Figure 2.9 to 2.11 In bakery design, the colour scheme plays an essential role and several elements should be considered
图 2.9~ 图 2.11 在烘焙坊设计中，配色至关重要，需要考虑多个要素
Figure 2.12, figure 2.3 Making full use of natural light and considering the effect of artificial light are two points in lighting design
图 2.12、图 2.13 充分利用自然光线、慎重考虑人造光源效果是灯光设计的关键之处
Figure 2.14 to 2.16 The refined paintings on the wall create a distinctive effect
图 2.14 ~ 图 2.16 精心设计的装饰画增添了独特的韵味

Bulka Cafe and Bakery

布尔卡咖啡烘焙坊 Moscow, Russia
俄罗斯，莫斯科

A summer house in Gorky Park
高尔基公园里的夏日小屋

Basic information　基本信息

Design/ 设计：Crosby Studios
(Chief designer/ 主设计师：Harry Nuriev)
Landscape design/ 景观设计：Anna Andreeva
Area/ 面积：228 m²
Photography/ 摄影：Evgeny Evgrafov

Key materials　主要材料

Wood
木材

Designed by Crosby Studios, Bulka Cafe and Bakery is located in the Gorky Park in Moscow in the middle of the extensive gardens along the Moskva River. Thus the greenery theme of the park is projected into the interior design through various details letting a bit of sunny summer into short overcast days and long winter nights of Moscow.

由 Crosby Studios 设计的布尔卡咖啡烘焙坊位于莫斯科的高尔基公园，藏在莫斯科河畔茂密的花园之中。因此，公园的绿化主题被投射到了室内设计之中，为莫斯科的阴天和漫漫冬夜带来了阳光明媚的夏日。

The floor plan is very simple and symmetrical and it can be easily transformed to meet various format requirements being either formal or casual.

The wide open space is divided into zones by industrial-looking shelvings decorated with potted plants and fresh vegetables waiting to be picked up by the chef for cooking.

室内空间的布局十分简单，呈对称式，并可以轻易地转化为不同的形式，满足各种正式、非正式的场景需求。

宽敞的开放空间被外观具有工业感的铁架分隔成不同的区域，铁架上装饰着盆栽植物和新鲜的蔬菜，这些蔬菜可供主厨随时采摘。

1. Main entrance
2. Bar
3. Main hall
4. Kitchen
5. Terrace
6. WC

1. 主入口
2. 吧台
3. 主就餐区
4. 厨房
5. 露台
6. 卫生间

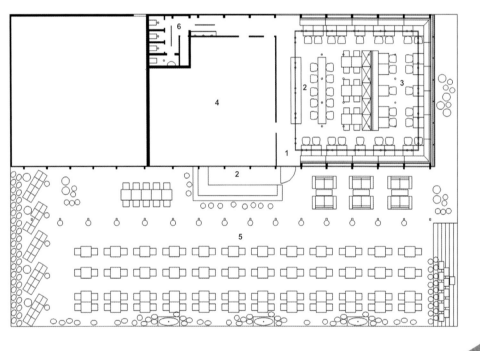

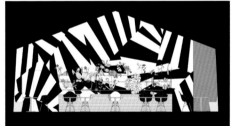

The main task was to create the most baking restaurant with the least baking interiors avoiding by implication all the associations with a traditional bakery like salvaged wood, vintage furniture, confiture jars, etc., and the architects came up with the idea of a summer house in the country but built for a young modern couple.

设计的主要任务是打造一个与众不同的烘焙餐厅，远离传统烘焙坊的打捞木材、复古家具、果酱罐等元素。最后，建筑师想出了"乡村夏日小屋"的点子，但是这个小屋要多加入一些年轻和现代的元素。

The bench stretching along the perimeter of the venue is custom made at Crosby Studios Workshop and painted jade colour just like the bar stools and the overhead lights. The tables just like the wooden floor are painted light grey.

贯穿整个周界的长椅由 Crosby Studios Workshop 特别设计，长椅被漆成碧玉色，与吧台高脚凳和头顶的灯架相互呼应。餐桌与地面都漆成了浅灰色。

Tableware and cutlery are stored in the open bar stand among viennoiserie and again potted plants. The back wall, the only one without windows, is painted in graphic dazzling pattern catching attention from outside through the other three glass walls of the pavilion.

The lighting fixture spreading 10 by 10 metres all around the venue is made of reading-lamps and pending potted plants repeating the pattern set by the shelvings.

餐具被储藏在外露的吧台上,与甜面包和盆栽一起。作为唯一一面没有窗户的墙壁,背景墙上绘制了炫目的图形,透过玻璃窗吸引着外面行人的目光。

跨越10x10米空间的灯具由台灯和悬挂的植栽组成,重复了铁架的花样。

Regular water pipes and fixtures just like one would use for outdoors water outlet in the country form the three faucets along the single wide sink made of tin ventilation channel. Solid wood floors are stained with grey shaded oil.

洗手池由废弃的锡制通风道制成,水池上方的水龙头和管道看起来就像乡村户外的水龙头和管道一样。实木地板上涂抹着灰色的清漆。

LOISIR

洛伊斯烘焙坊 Seoul, Korea
韩国，首尔

An achromatic colour space but with fascinating life
素色空间，精彩生活

Basic information 基本信息
Design/设计：Nordic Bros. Design Community
Area/面积：80m² (ground floor/一楼) + 90m² (first floor/二楼) +67m² (terrace/露台)
Photography/摄影：Nordic Bros. Design Community
Completion/完工时间：2014.8

Key materials 主要材料
Terracotta, wood (floor)
Brick, paint (wall)
Paint (ceiling)
瓷砖、木地板（地面）
砖、涂料（墙面）
涂料（天花板）

Loisir is a baking studio & boutique cafe owned and operated by patissier Sookyung Kim about afternoon tea and cake. Loisir got inspiration from the 'time seeking solitude of life' and will interpret and show desserts in France and Japan after Loisir's own fashion.

The client wanted Loisir's own sensibility interpreted by the designer could be expressed in a space just like the 'space achromatic coloured, unexposed but fascinating itself' and she wanted to commune with the designer through various themes.

洛伊斯是一家烘焙工作室兼精品咖啡屋，由糕点师金秀卿执掌，专营下午茶和蛋糕。洛伊斯的设计灵感来自于"追求独处的时光"，它用自己独特的风格来诠释法式甜点和日式甜点。

客户希望设计师将洛伊斯呈现为"一个素色空间，隐蔽而令人着迷"，为此，她与设计师探讨了许多个设计主题。

During the process of understanding the client, the designer deduced the direction towards the 'space changeable each season' and focused on allowing a margin at important points of the space while concentrating on the layout such as the utilisation of interior and exterior spaces separated into two stories and minimisation of the traffic line.

Before selecting materials and colours, the client asked to preserve the appearance (used bricks) as much as possible upon request by the landlord, and the designer put emphasis on the use of unsmooth finish materials such as terracotta and painting of the exposure state to fit in the used bricks.

在了解客户的过程中,设计师将设计方向定位为"随着四季变化的空间",决定在重要的空间节点留白,同时注重室内外空间的划分、交通动线的极简化等布局概念。

在选择建筑材料和色彩之前,客户要求尽量保留砖墙的原样,因此设计师将重点放在了陶瓦等不平滑的装饰材料上,并且将涂料直接涂抹在裸露的砖墙上。

Ground Floor Plan
1. Baking studio
2. Staff room
3. Storage
4. Kitchen
5. Toilet
6. Office
7. Boutique shop

一层平面图
1. 烘焙工作间
2. 员工休息室
3. 储藏间
4. 厨房
5. 洗手间
6. 办公室
7. 精品店

First floor inspired by an arcade and a corridor

To utilise merits of the previous environment (the house with an open yard and a quiet alley), it blurred the distinction between spaces and emphasised the continuity by arranging the gable roof structure consecutively in a row inside and outside.

While providing each table inside and outside with the individual view and leaving a margin for a line for lighting equipment, it used copper as the accent colour to express the space image that fell under narcissism.

二楼从拱廊和走廊中获得灵感

为了利用原有空间的特色（房屋有一个开放的庭院和一条幽静的小巷），设计模糊了空间的界限，通过室内外一系列的山墙屋顶结构突出了空间的连续感。

室内外的每个餐桌都享有独立的风景，中间的留白空间安装着照明设施。被铜色点缀的空间散发出一种自我陶醉的感觉。

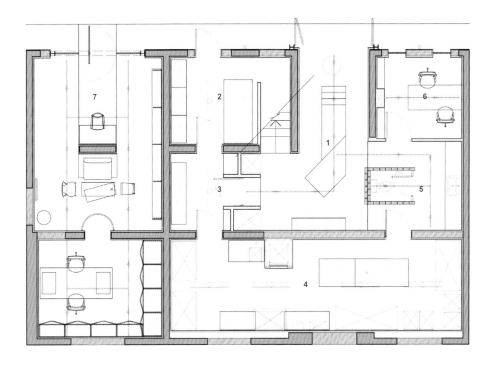

Ground floor substantial in the utilisation of space

After combining two previous studio apartments into one and placing the baking studio and the stairs going to the cafe upstairs in the centre, the staff room, storage, main kitchen, toilet, and office enclosed them in the form of ' え 'to minimise the traffic line and fully utilise the space.

一楼的设计以空间利用为主

设计将两个小型公寓合并,把烘焙工作室和通往咖啡厅的楼梯设在中央,四周环绕着员工休息室、储藏室、主厨房、洗手间和办公室,组成了日语中的"え"字型,简化了交通动线,充分利用了整个空间。

First Floor Plan
1. Entrance
2. Terrace
3. Storage
4. Bar
5. Cafe
6. Stair
7. Toilet

二层平面图
1. 入口
2. 露台
3. 储藏间
4. 吧台
5. 咖啡间
6. 楼梯
7. 卫生间

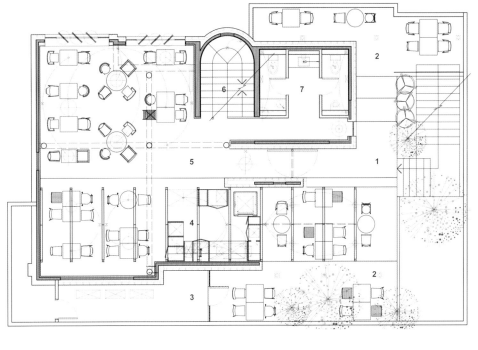

hey ju

"嘿！宇"烘焙工坊 Cheonan-si, Korea
韩国，天安市

A baking studio focusing on colours
体现特有色彩的烘焙工作室

Basic information 基本信息
Design/ 设计：Nordic Bros. Design Community
Area/ 面积：103m²
Photography/ 摄影：Nordic Bros. Design Community
Completion/ 完工时间：2014.12

Key materials 主要材料
Tile, white epoxy (floor); paint (wall, ceiling)
瓷砖、白色环氧树脂(地面)；涂料(墙壁、天花板)

The 'hey ju' dessert cafe & baking studio of the patissier Jung-Ju Kim is located in the culture and cafe specialised street in Buldang-dong, Cheonan-si, Korea.

The client wanted the baking studio and cafe to have its own colours in one space among the various shops lined up in the street and make them become a space that focuses on the communication with visitors as a cafe where they can enjoy brunch, dessert, and tea together with a baking class based on his preference of achromatic colour.

糕点师金正宇的"嘿！宇"甜点咖啡兼烘焙工作室位于韩国天安市一条以文化和咖啡著名的街道上。

客户希望烘焙工作室和咖啡厅在店铺林立的街道上拥有自己特有的色彩。一方面，顾客可以尽情享用早午餐、甜点和茶点；另一方面，顾客也可以根据自己的色彩偏好参与到烘焙教学中。

Floor Plan
1. Entrance
2. Cafe
3. Baking studio
4. Service bar
5. Patissier kitchen
6. Powder & toilet

平面图
1. 入口
2. 咖啡间
3. 烘焙工作间
4. 服务吧台
5. 糕点制作厨房
6. 洗手间

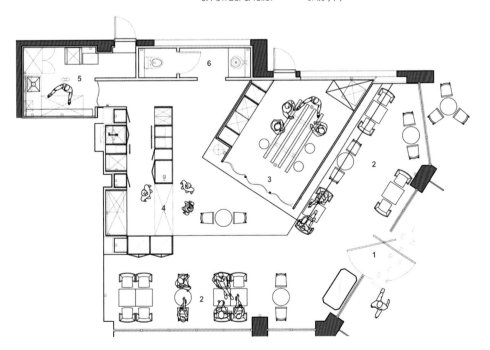

While concentrating on the layout where the cafe and the baking studio are put together, which was the biggest concern of the client, the designer drew a 'surreal' concept and planned the traffic line key to the service of the patissier around the baking studio.

While planning the baking studio and the cafe that offer different services although there is a close connection between them in one space, the designer had a conversation to understand the client and completed an atelier of the patissier Jung-Ju Kim which properly harmonised open and close.

考虑到咖啡厅和烘焙工作室合二为一的主题，设计师起草了一个"超现实"概念，根据糕点师在烘焙工作室里的服务规划了交通动线。

虽然处在同一个空间，但是烘焙工作室和咖啡厅所提供的服务是不同的，设计师通过对话了解了客户的需求，设计了一个可根据需要自由开闭的工作室。

Upper level

A space inside a space covered with a roof in a diagonal form is made in the centre of the space to place the baking studio, and the raised floor height (baking studio, service counter, kitchen, toilet) divides the boundary between the baking studio and the cafe.

上层

烘焙工作室被设在空间的中央，上面特别设计了一个斜屋顶，垫高的地面将烘焙工作室、服务台、厨房、洗手间与咖啡厅的空间隔开。

Lower level

A big neon sign is put on the exterior wall (in the front of the entrance) and the roof of the baking studio, and tables and chairs are placed around the glass wall where the whole street can be viewed to enjoy the street culture blurring the boundary between inside and outside.

下层

店面入口外墙和烘焙工作室的屋顶上各有一个巨大的霓虹灯招牌。餐桌椅都环绕着玻璃墙摆放，顾客可以对街道上的情景一览无余，室内外的界限已经变得模糊。

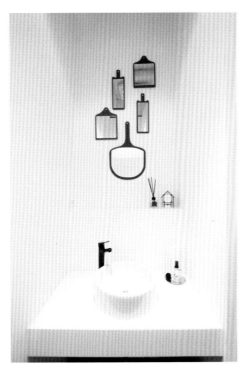

The White Gift Box - Aimé Patisserie

Aimé 烘焙坊 Shanghai, China
中国，上海

Unwrapping the white box
拆开白色礼盒

Basic information 基本信息

Design/ 设计 : Christina Luk, Mavis Li, Wesley Shu, Scott Baker/ LUKSTUDIO
Area/ 面积 : 63m²
Photography/ 摄影 : Peter Dixie for LOTAN Architectural Photography
Lighting consultant/ 灯光咨询 : German To for Lucent Lit Co., Ltd.
Completion/ 完工时间 : 2014.2

Key materials 主要材料

White display case
白色展示架

Aimé Pâtisserie, a new brand entering the Shanghai market, has positioned its flagship store in an elevated retail strip on Huai Hai Road. The chosen site is flanked by familiar coffee and donut franchises, and fronted by a city bus-stop. Across the street is the recently opened shopping mall flaunting an array of luxury brands. The design challenge of the store is to stand out from its immediate chaos and appeal to the clientele from the close-by gentrified neighbourhood. The strategy is to dress this newcomer up as a white present.

全新品牌 Aimé Patisserie 落户上海，选择了在淮海路的一片小台阶上建立旗舰店。它位于一个全新的大型高端购物中心对面，两旁尽是熟悉的咖啡店与油炸圈饼店，门前是一个繁忙的公共汽车站。这里人流众多但竞争激烈，不少商铺都在改善装潢、提高品质，所以它面对的第一个挑战就是在这里突围而出、吸引顾客，而设计方案就是让这家新店化身为一个全白色的礼品盒。

Floor plan
1. Main entrance
2. Window display
3. Bar seating
4. Waiting area
5. Service area
6. Display counter
7. Staff zone
8. Back of house kitchen
9. Shadow art logo
10. Packaging counter

平面图
1. 主入口
2. 橱窗
3. 吧台就餐区
4. 等候区
5. 服务区
6. 展示柜台
7. 员工区
8. 厨房
9. 艺术标识
10. 包装区

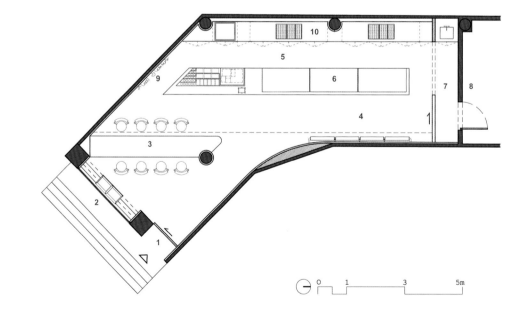

Beyond the 4.5m retail storefront lies a turning L-shape layout divided into two zones: the foyer with bar seating and the display counter at the back. With a lower ceiling, the former compresses views of the latter, yet frames the illuminated feature wall to capture the curious minds. Every step forward heightens the discovery of the playful interior where the 9m long counter showcases macarons and other goodies in perfect order.

在4.5米宽的门面以内，是个呈L形的空间，分为两个部分：门厅有一张供顾客享用美食的吧台，里面是个陈列产品的柜台。前者层高略低一点，虽然稍微压缩了后者的视觉，但正因为如此，设计师把顾客的视线集中在发亮的主墙上。当你往放满马卡龙与其他精美甜品的9米长柜台走过去，就越发觉得店里的细节可供玩味。

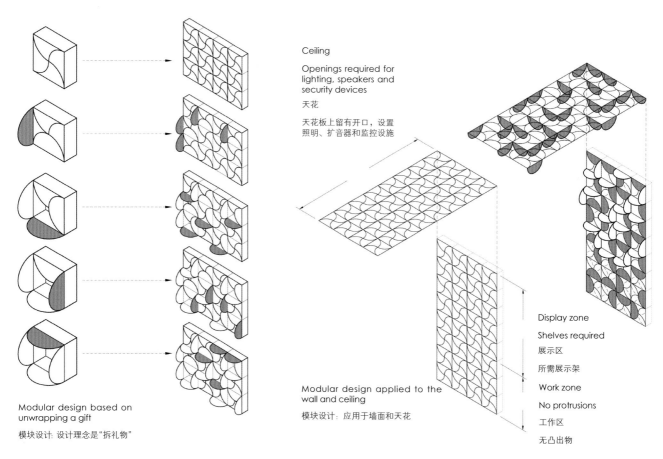

Ceiling

Openings required for lighting, speakers and security devices

天花

天花板上留有开口，设置照明、扩音器和监控设施

Modular design based on unwrapping a gift

模块设计：设计理念是"拆礼物"

Modular design applied to the wall and ceiling

模块设计：应用于墙面和天花

Display zone

Shelves required

展示区

所需展示架

Work zone

No protrusions

工作区

无凸出物

The illuminated feature wall composed of stacked gift boxes at various states of opening draws one's eyes up to the ceiling were the whimsical pattern continues. The versatile modular system allows for necessary wall display shelves, and ceiling openings for spotlights, speakers, and security devices.

In light of the neutral palette evoking a gallery ambiance, the branding wall is created with an artistic approach where aluminium bands are twirled and bent to cast calligraphic shadow. A visit to this white gift box should be a delight to the eyes as much as to the sweet tooth.

主墙的设计就好像排列着一个又一个的礼品盒，都以不同的形态打开着，吸引你往上看尽各种如梦似幻的图案。这个灵活的装置不但美观地展示各种产品，还能收纳聚光灯、音响系统和保安系统等设备。

由于店面以白色为主调，带来一点画廊的气质。我们希望以一个比较艺术的手法去创作陈列柜台后的品牌墙。设计师跟一位修读美术系的学生合作，把一组铝质条子扭曲，让它们投射在墙上的影子呈现品牌名称。这个充满美感的视觉效果，令到访的顾客们除了被甜美的食品迷倒，也会对这个"白礼盒"留下一个精致的印象。

Bakery Marie Antoinette

玛丽皇后烘焙坊 Bogotá, Colombia
哥伦比亚，波哥大

Romance and femininity arising from here
这里散发着浪漫的女性气息

Basic information 基本信息
Design/ 设计：David Pinilla
Area/ 面积：108 m²
Photography/ 摄影：Juan Castro
Completion/ 完工时间：2014

Key materials 主要材料
All materials are locally created
所有材料均由本地生产

This space recreates and evokes the fantasy of a queen who lives in our imagination. The expression of Versailles and Marie Antoinette's grace is reflected through shapes, colours and textures of the furniture, and everyday objects, recreating a unique history and making something of the past now present.

这个空间重现了那位活在我们想象中的皇后的梦幻传奇。凡尔赛宫和玛丽·安托瓦内特皇后的优雅投射在家具的造型、色彩和质感以及日常用品上，重现了那段独特的历史，将昨日的传奇呈现在当下。

At the International Centre of Bogota, near the National Museum, resides the bakery. The approximate measurement is 108m²; it is distributed in two floors, with a capacity for 40 seated guests. All materials and manufacturing were locally created in the design studio.

Based on the above, the site offers an experience that allows one to feel, hear, taste and perceive the entire universe of the young queen's decadent lifestyle. The space evokes romance and femininity all while remaining classic and refined.

本店位于波哥大国际中心，紧邻国家博物馆，总面积约108平方米。烘焙坊分为两层，总共可提供40个座位。所有材料和生产过程都在设计工作室当地完成。

以此为基础，烘焙坊为顾客提供了从视觉、听觉到味觉的全方位体验，完美呈现了这位年轻皇后的生活方式。整个空间充满了浪漫的女性气息，同时又不失精致与典雅。

Raining chocolate and mint walls blend with the sweet smells from the kitchen. Glass cabinets display dainty almond croissants, chocolate cakes and pastel cupcakes of all flavours. An exposed kitchen allows customers the pleasure to view a unique baking process full of dedication, talent and passion. "Everything to discover lies across the open window."

巧克力雨点墙和薄荷绿色的墙壁与厨房的香气融为一体。玻璃壁橱里展示着美味的杏仁牛角包、巧克力蛋糕和各种口味的彩色纸杯蛋糕。开放式厨房让顾客可以观看充满了奉献精神、技巧和热情的整个烘焙过程。"窗口的另一端有着无穷的发现。"

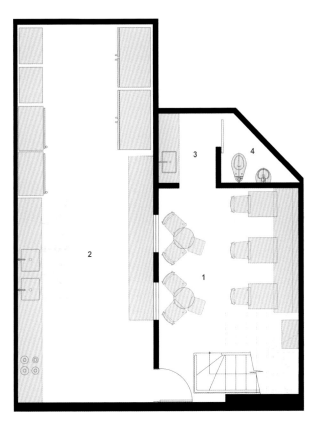

Lower Floor Plan
1. Dining area
2. Kitchen
3. Vanity table
4. W.C.

地下层平面图
1. 用餐区
2. 厨房
3. 梳妆台
4. 洗手间

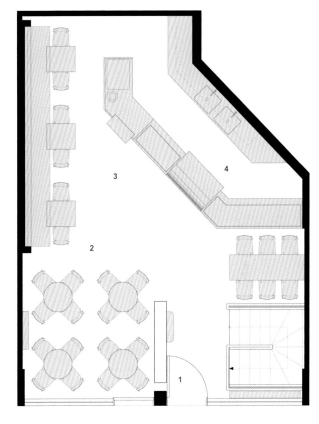

Ground Floor Plan
1. Entrance
2. Dining area
3. Counter
4. Kitchen

一层平面图
1. 入口
2. 用餐区
3. 柜台
4. 厨房

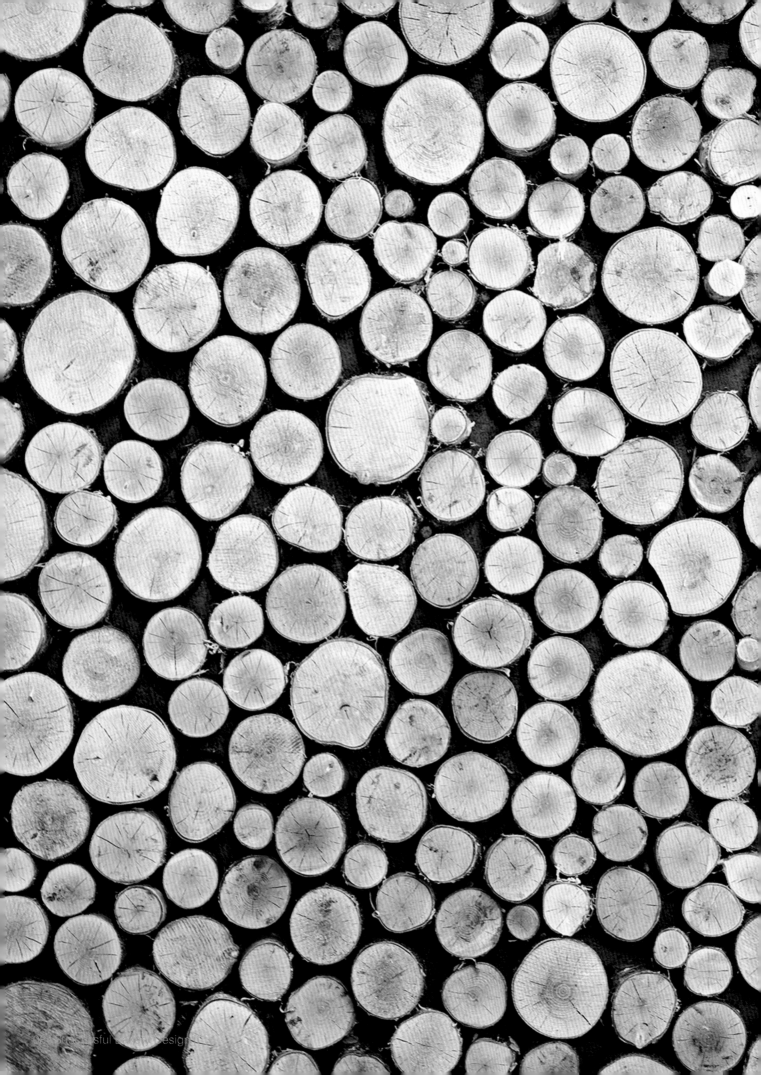

Parken Bakery

帕肯烘焙坊 Oslo, Norway
挪威，奥斯陆

A 'green lung' in an old building
旧建筑中的"绿肺"

Basic information | 基本信息
Design/ 设计 : LINK arkitektur
Design team/ 设计团队 : Daniela Colli architect
Photography/ 摄影 : Wenche Hoel-Knai
Completion/ 完工时间 : 2013

Key materials | 主要材料
Wood
木材

Construction work on new buildings and renovations has been ongoing in the Science Park for many years, and LINK arkitektur won a competition for the task of creating a new bakery café in the oldest building. The concept is based on the idea of creating an internal 'green lung' as a natural meeting place for staff and visitors.

科学公园一带的新建工程和翻修工程已经进行了多年，LINK arkitektur 此次通过设计竞赛获得了一项在该地区最古老的建筑中设计一家烘焙咖啡馆的任务。设计的概念以打造室内"绿肺"为基础，力求为楼内的员工和访客打造一个天然的会面场所。

Floor Plan
1. Bar counter
2. Dining area
3. Private meeting
4. W.C.

平面图
1. 吧台
2. 用餐区
3. 私人区
4. 洗手间

Strong use of colour, wallpaper and furnishings are largely a contrast to the building in general. It is an attractive and pleasant place, in which it is comfortable both to sit and work and to take a well-deserved break. The furnishings are varied and invite socialisation, while it also provides opportunities for those who prefer to sit alone.

大量的运用色彩、壁纸和装饰在整体上与建筑形成了鲜明的对比。这里充满了吸引力，可以坐下来舒服的工作，也可以尽情地休息片刻。室内装饰多姿多彩，既鼓励人们进行社交，又为喜欢独处的人提供了安静的角落。

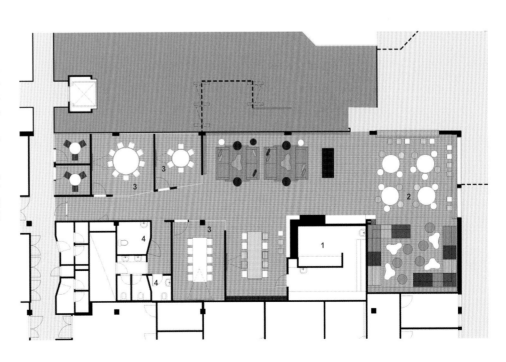

The bakery is equipped with meeting rooms that can be booked at any time. This combination is innovative and we can see that this could be a concept for the future.

烘焙坊内设有会客室，可以随时预订。这种组合十分新颖，为未来烘焙坊的发展提供了新的思路。

The Cake

蛋糕店   Kiev, Ukraine
乌克兰，基辅

Simple yet fashionable in colour scheme
配色突显简约与时尚

Basic information | 基本信息

Design/ 设计：Slava Balbek, Nadya Chabanny
Area/ 面积：150m²
Photography/ 摄影：Andrey Bezuglov,
Slava Balbek
Completion/ 完工时间：2014.11

Key materials | 主要材料

concrete tile (floor)
混凝土砖（地面）

The architects disputed about the name for the design longer than worked on it, which took record-breaking shortest term: four months from first sketches till open day rehearsal. However, they have eventually reconciled with calling it after a music genre – 'indie' design.

The diversity of images and forms presented in the 80m² main hall has only been achieved due to the fact that most of the design elements have been handcrafted.

本次设计从第一张草图到开张预演只用了4个月的时间，而建筑师针对本次设计的命名讨论时间甚至超过了设计工作时间。最后，他们决定参考"独立音乐"，将设计命名为"独立设计"。

各种各样的图形和图像呈现在80平方米的主厅内，大多数设计元素都是纯手工制作的。

Sun light walks in through massive windows, fills up the whole space and reaches the opposite glass walls adding contrast to the beautifully carved bar wooden panels on its way.

阳光从宽大的窗口进入室内，填满了整个空间，直达对面的玻璃墙，为雕刻精美的吧台木挡板增添了对比感。

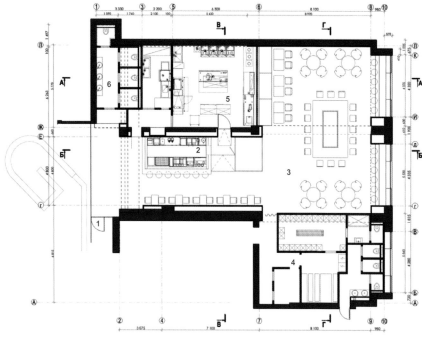

Floor Plan
1. Entrance
2. Bar
3. Dining area
4. Staff room
5. Kitchen
6. W.C.

平面图
1. 入口
2. 吧台
3. 用餐区
4. 员工休息室
5. 厨房
6. 洗手间

On a pale and colour-muted background of the interior, a glossy pink sculpture looks extremely contrast – just like a glazed cake on a white plate.

在色彩清淡的背景下,富有光泽的粉红色雕塑品显得异常突出,就像是白盘子里一块上糖釉的蛋糕。

Two height seating areas have their own dedicated colour and tactile symbol. Rustic wooden high surface with felt thin bar-chairs standing on metallic racks neighbours with velvet fabric of cosy bottle-green couches and sofas covered with warm shade of grey.

两个高低不同的座位区分别被冠以独立的色彩和材质。一边是做旧的木板围桌，配置毛毡面金属腿的吧台凳；一边是深绿色天鹅绒面料沙发椅和浅灰色沙发。

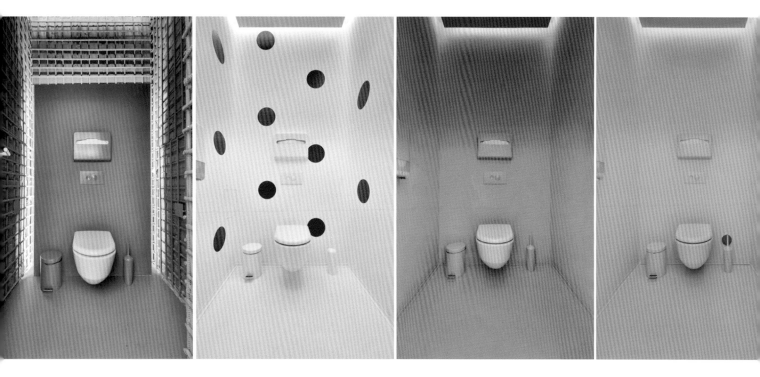

Consistency in contrasts has been demonstrated in mono-colour toilets: pink, deep-green, yellow, reminding that simple solutions might yet look very unusual.

纯色卫生间设计融合了对比与统一,有粉色、深绿、黄色等颜色。简单的设计有时也能让人耳目一新。

In contrast to the piles of cakes of ideal geometrical form installed behind the transparent walls of the kitchen, every floor concrete tile has its wrinkle as each of 10,000 of them has been handcrafted. Yet altogether they make up harmonised puzzled surface.

与厨房玻璃墙后面所制作的造型完美的蛋糕相比，每一块混凝土地砖的纹理都略不相同，因为这10,000块地砖全部是手工制作的。这些地砖组合起来，共同构成了和谐而奇妙的地面。

Royal Bakery

皇家烘焙坊 Kiev, Ukraine
乌克兰，基辅

Appetising space of warm colours
美味的暖色系空间

Basic information 基本信息
Design/ 设计 : Kozyrniy Design
Design team/ 设计团队 : Aleksandr Yudin, Vladimir Yudin, Sergey Proshutya
Area/ 面积 : 120 m²
Photography/ 摄影 : Dmitry Sandratsky
Completion/ 完工时间 : 2013

Key materials 主要材料
Lightwood (bar), wood (wall)
轻木（吧台）、木板（墙壁）

The Royal Bakery was established in 1906. It is situated in a picturesque place on the bank of the river Dnipro called Obolon' seafront. The location abounds in natural colours and textures; the place is peaceful and cosy at any time of the day.

皇家烘焙坊建立于 1906 年，坐落在风景如画的迪尼普河畔。整个空间充盈着自然的色彩和质感，无论在任何时间，这里都十分舒适而平和。

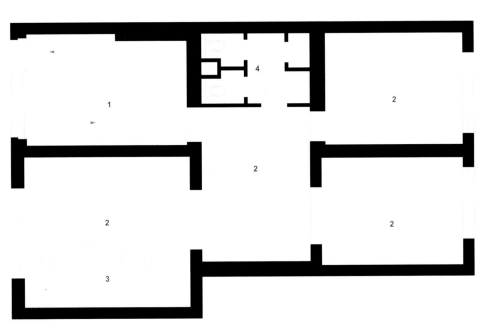

Rectangular shape of the shop enabled the architects to design three zones – entrance, hall and bar, staff and storage area in the back. The commission was to create a coherent interior design for a shop. Key features of the shop are a custom-designed display shelving and a counter. Another main feature used in all interior is wooden wall creating warm hues.

长方形空间布局使烘焙坊可以分为3个区域——入口区、大厅吧台区、员工仓储区。项目的目标是为烘焙坊打造一个连贯的室内设计。烘焙坊的主要特征是定制的展架和柜台。室内的木板墙面营造出温暖的氛围。

Floor Plan 平面图
1. Entrance 1. 入口
2. Hall 2. 大厅
3. Bar 3. 吧台
4. W.C. 4. 洗手间

Having settled themselves in cosy armchairs, guests enjoy fresh bakery, fragrant coffee and, certainly, relaxed conversation.

坐在舒适的扶手椅上,顾客能一边享用新鲜的烘焙产品和香醇的咖啡,一边进行轻松的交谈。

All the areas are kept in the same style, and the entrance with red sofas are more comfortable for children. There is a blackboard wall for announcing special offers and stainless steel furnace for baking fresh buns and delicious cookies on site.

所有区域都保持着统一的风格,入口的红色沙发更适合儿童使用。黑板墙面书写着每日特卖,不锈钢烤炉每天都烘焙出新鲜的小圆面包和美味的饼干。

Patisserie Pan y Pasteles

西班牙马德里 Pan y Pasteles 甜品店　Madrid, Spain
西班牙，马德里

A pink space, a sweet dream
一个粉色的空间，一个甜美的梦

Basic information 基本信息
Design/ 设计：Ideo Arquitectura
Area/ 面积：55m²
Photography/ 摄影：Evgeny Evgrafov
Completion/ 完成时间：2015

Key materials 主要材料
12,000 pink-painted wooden sticks (ceiling), micro cement (floor)
12,000 根粉色木板条（天花板）、超细水泥（地面）

The project consists of a new design of a third bakery in Madrid which makes bread and cakes. The client believes that every bakery should be unique and different to the other ones and the only specific design request is the use of their coorporate colour, the magenta.

Pan y Pasteles 甜品店制作面包和糕点，这个项目是该品牌在马德里的第三家店面。店主认为每个店铺都应该是独一无二的，跟其他的店铺有所不同，唯一的设计要求是：使用该品牌的标志色——品红色。

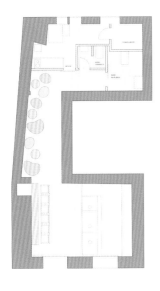 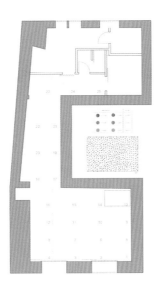

In order to achieve a contemporary design, which was the initial objective, they had to find an element with a strong character that would compete with the consistency of the 150-year-old framed walls without eclipsing them. Therefore they created an artistic installation: more than 12,000 wooden magenta sticks that are hanging from the ceiling. This installation attracks everyone's looks!

为营造出现代设计的气息（这也是最初确定的设计目标），设计师要找到一种极具特色的元素，用来装饰有着150年悠久历史的墙面，同时要保证不会腐蚀墙面。为此，设计师特别打造了一种艺术装置：12,000根木板条从天花板上悬垂下来，绝对吸引眼球！

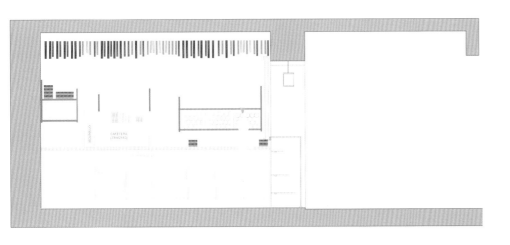

The architect and designer, Virginia del Barco, has designed the lighting, a part of the furniture, like the chairs, the stools, the shelves, the bartop, the whiteboards, and the light boxes in the façades. On the pictures, you can appreciate the different details of the claddings, as well as the high-quality micro cement pavements which give the space a refined elegance.

设计师弗吉尼亚·德·巴尔科还贡献了照明设计,作为家具设计的一部分,包括椅子、吧凳、货架、吧台顶面、白色书写板以及外立面上的灯箱等。从这里呈现的照片上可以看到覆层材料的细节表现,包括超细水泥的地面,让空间显得精致、典雅。

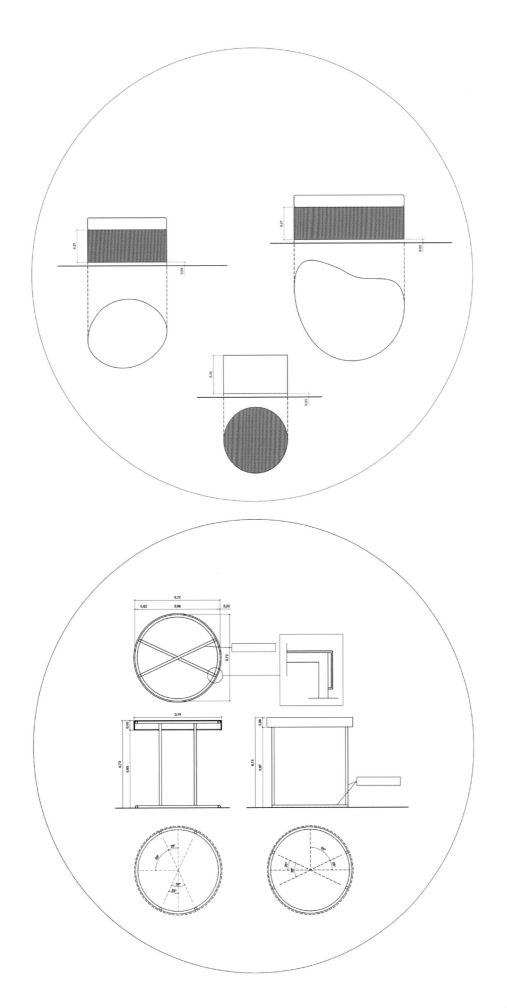

Vyta Santa Margherita Bakery

圣玛格丽塔烘焙坊　Milan, Italy
　　　　　　　　　　　意大利，米兰

Enjoy the original food philosophy in Central Milan Station
享受米兰中央车站的原创美食哲学

Basic information　基本信息
Design/ 设计 : Collidaniel architetto
Design team/ 设计团队：Daniela Colli architect
Area/ 面积 : 180 m²
Photography/ 摄影 : Matteo Piazza – Milan
Completion/ 完工时间 : 2013.11

Key materials　主要材料
Pearl sicilian marble slabs (inside), original decoration in mosaic and marble (outside); wall: glossy black polymer; ceiling: black painted plasterboard and curved oak wood slats
珠光西西里大理石板（内）、马赛克和大理石原创装饰（外）；墙面：光泽黑色聚合材料；天花板：黑色涂料石膏板和弯曲橡木板条

Vyta Santa Margherita offers the oldest and most traditional products, bread and wine, in one of the most representative places in the city, Central Milan Station, the symbol of Milan's hectic urban life.

"Through simple products offered by Nature, such as water, wheat and fire, thanks to man's expert hand, patience and creativity, forms and savours, aromas and flavours have been created for millennia, giving birth to bread and wine, ancient and modern nourishment for humanity."

This food philosophy was the starting point that inspired the architectural concept. A contemporary look has been reformulated for the most 'minimal' products on our tables. It originates from a restraint design and an innovative, cool elegance, the result being a sophisticated minimalism and a formal reduction to the essential.

圣玛格丽塔烘焙坊专营最古老、最传统的美食、面包和红酒，它坐落在最能代表米兰繁华的城市生活的地点——米兰中央车站。

"人类用灵巧的双手、耐心和创造力，用大自然赋予我们的简单材料（水、小麦、火等）创造出美味，制作出面包和红酒，为人类提供了最古老又最现代的营养美食。"

这一美食哲学就是本项目设计概念的出发点。现代的外观被重新设计，以适应我们餐桌上最"简单"的食品。简约的设计和新颖的优雅感最终形成了精致的极简主义设计，回归到了生活的本真。

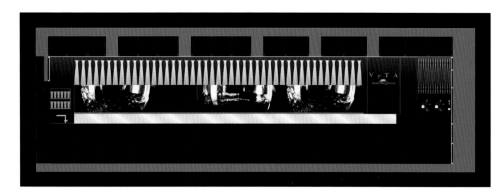

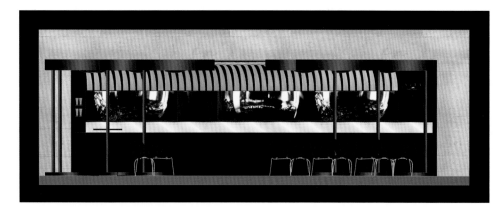

The project features contrasting materials and colours: oak and corian as representatives of tradition and innovation, an integration of nature and artifice. The juxtaposition of soft oak and black declined in its various material aspects creates an exclusive, theatrical environment, where the warmth of the natural texture is enhanced by the contrast with glossy black surfaces. These come up as corian for the counter and black polymer for all the vertical panels that fold the space like in a treasure chest.

The counter, with its soft curved line, is a fluid shape emphasised by a roof made of curved slats, one of the most significant components of the setting, due to the shape and size of its natural oakwood planks that evoke the interweaving of traditional bread baskets.

The wall of the back counter, usually just a functional space, becomes a new form of entertainment through a video wall that reproduces a perpetual movie in slow motion that celebrates the ingredients of the products Vyta-Santa Margherita such as water, wine, olive oil, wheat, flour, mozzarella, tomatoes and fire.

项目所运用的材料和色彩都极具对比效果：橡木和可丽耐人造大理石分别代表着传统与创新、天然与人工。柔和的橡木与黑色的材料并置在一起，营造出一种具有戏剧感的独特环境，富有光泽的黑色表面进一步凸显了天然材料的温暖质感。可丽耐台面和黑色聚合材料将整个空间包裹起来，就像一只珍宝箱。

柜台采用柔和的曲线造型，天花板上的弯曲板条进一步突出它的流畅感。作为整个场景中最重要的构成元素，天然橡木板条的形状和尺寸都令人容易想起传统的面包篮子。

柜台后面的墙壁通常只是一个功能空间，但是本项目利用视频墙引入了全新的娱乐形式。缓慢变化的影像歌颂了烘焙坊所选用的材料，例如水、红酒、橄榄油、小麦、面粉、马苏里拉奶酪、番茄和火。

Floor Plan
1. Dining area
2. Bar
3. W.C.

平面图
1. 用餐区
2. 吧台
3. 洗手间

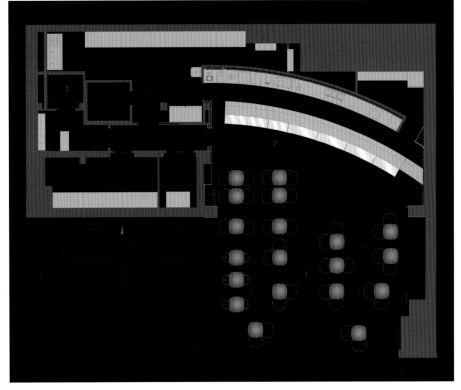

The light system contributes to soft and intimate atmospheres: it diffusely radiates on the counter and enhances the wooden roof by indirect fitting of recessed metal halide light sources Minimal by IGuzzini. It is an eye-catcher on the black wall where the bottles of wine and their history are celebrated like sculptures by the backlit decoration; realised by milling thickness of slabs, it gets highly technological to celebrate bread and its derivatives through a cluster of LEDs.

照明系统营造出柔和而亲切的氛围：灯光呈辐射状照射在柜台上；金属卤素灯（IGuzzini 的 Minimal 系列）的间接照明提升了木天花板的设计感。黑色墙壁上，加工的木板条以其独特的造型烘托着红酒瓶，十分引人注目。LED 灯群的设计让面包及其衍生产品作为主角，呈现在顾客眼前。

116 Successful Bakery Design II

The monumental atrium station has been brought down to a human architectural scale, through the creation of technological and custom-made umbrellas; made of black metal they are equipped with infrared light with low energy consumption for heating, LED lighting, sound system and an electrical outlet to recharge smartphones and tablets, so that the space has a less monumental and more intimate look and at the same time the ensemble evokes the ancient rite of eating together, a less common practice nowadays, but increasingly necessary in the third millennium's life.

特别定制的黑色金属伞把宏伟的中庭车站分割成人性化的建筑尺度。这些伞配有低能耗的加热红外光、LED照明、音效系统和电插座（方便手机和平板电脑充电），让整个空间变得不那么宏伟，而是多了一丝亲切感。整体设计唤醒了人们聚在一起就餐的传统习俗，这正是我们现代社会所缺少并且急需的。

Zhengzhou Industrial Style Bakery

郑州工业风烘焙店　Zhengzhou, China
　　　　　　　　　中国，郑州

The European industrial style embodying natural and genuine feelings
欧洲工业风所隐含的质朴与纯真

Basic information 基本信息
Design/ 设计：Er Heyong Space Design Group/ 二合永空间设计事务所
Design team/ 设计团队：Cao Gang, Yan Yanan/ 曹刚、阎亚男
Area/ 面积：90 m²
Completion/ 完工时间：2014

Key materials 主要材料
Square steel, aluminium plate, log, concrete
方钢、铝板、原木、混凝土

The only character of commercial stores design is 'product'. The design takes European industrial style as prototype and makes full use of the materials' features, such as reinforced concrete's simplicity and purity, to set off the products' fresh, delicious and healthy characters.

商业门店类的设计主角只有一个，就是"产品"。本案设计原型来源于欧洲工业风格，在材质运用上充分利用了钢筋混凝土等材质的质朴、本真，来衬托出要售卖产品的新鲜、美味和健康。

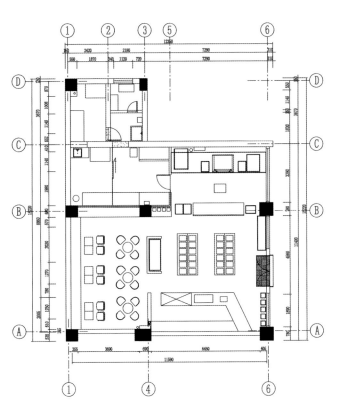

With 'six senses' as a starting point, the emotional setting examines the design associated to the shop's molding materials from the customers' perspective.

情感营造方面以人的"六感"为出发点，从顾客角度审视整个店面造型材料所能联想到的场景感。

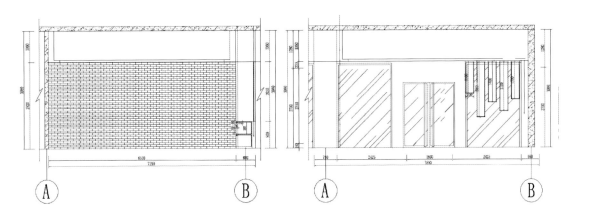

Successful Bakery Design II 121

In some spaces, the design makes the most of people's curiosity to plan the circulation, complemented with the supplementary lighting design, audio system and temperature control to lead customers to choose and buy their products in a desired way. At the same time, customers won't feel being forced and will purchase in a relaxing atmosphere without pressure, which inspires their potential purchasing power considerably.

在局部的空间里充分利用人的好奇感来对动线进行规划，再配以灯光、音响、温度的辅助设计让顾客既能按照设计预想的线路进行选购，同时顾客又无被逼迫感，在轻松无压的气氛中消费，从而激发顾客的潜在购买力。

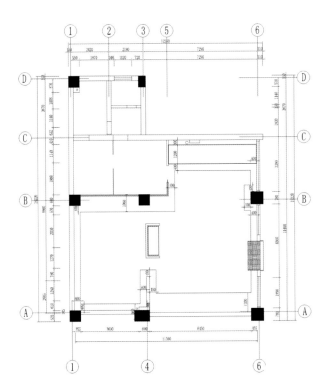
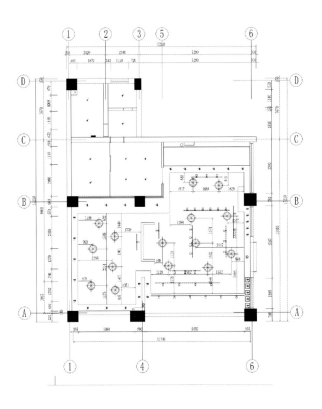

Kao Nekad Restaurant & Bakery

旧时光烘焙餐厅 Novi Sad, Serbia
塞尔维亚，诺维萨德

Extension of the traditional bakery concept and bringing back childhood memories

拓展传统的烘焙概念，带回童年记忆

Basic information 基本信息
Design/ 设计 : Synthesis Quatro / Slavica Djokovic /Aleksandar Osei-Lartey
Photography/ 摄影 : Dragana Mirkovic
Completion/ 完工时间 : 2014

Key materials 主要材料
Wood
木材

Domestic food restaurant 'Kao nekad' is an extension of the basic bakery concept of 'Kao nekad' bakery chain, since the company wanted to extend the offer for their clients in Novi Sad, Serbia.

The idea behind the extension was to bring nature inside the restaurant, alongside the basic concept of bringing back the memories from the childhood as the name of the company says (Kao nekad - Like in the old times).

旧时光烘焙餐厅对旧时光连锁烘焙坊的基本烘焙概念进行了拓展，因为公司希望在塞尔维亚的诺维萨德为客户们带来进一步的美食体验。

本次概念拓展背后的理念是在餐厅中体现烘焙的品质，同时，正如公司的名字"旧时光"一样，为顾客们带回童年的记忆。

Space is divided into three segments, dining, food and bakery product orders and self-service (salad bar, freshly squeezed juices and cakes). Segments are highlighted by the different light intensity and movement through the space is directed by the light and furniture.

Design and materials used had to be simple and basic, so reuse of the kitchen equipment and accessories has been the main and inexhaustible inspiration hand in hand with the greenery usage.

空间被分为3个部分：就餐区、食品点餐区和自助区（沙拉吧、鲜榨果汁和蛋糕），以不同的光照强度来区分，直接用灯光和家具来引导人们在空间内的移动。

空间的设计与材料必须简单基本，厨房设备和配饰的重新利用以及绿化设计成为了取之不尽的灵感来源。

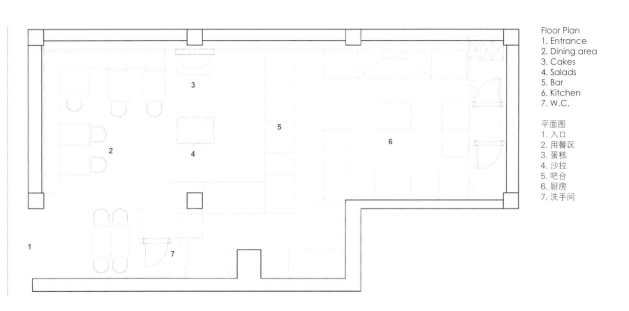

Floor Plan
1. Entrance
2. Dining area
3. Cakes
4. Salads
5. Bar
6. Kitchen
7. W.C.

平面图
1. 入口
2. 用餐区
3. 蛋糕
4. 沙拉
5. 吧台
6. 厨房
7. 洗手间

Thanks to the large windows, it wasn't hard to bring nature in. Potted vegetables and herbs hang from the ceilings and line the shelves. In particular, the central shelf puts plants at the heart of the design, where they may be admired by visitors and used by the bakers. The piece of furniture also divides the room so that the space feels cosy from all corners, enriching the restaurant with the feeling of being within grandmother's dining room. 'Kao nekad' bakery manages to create a different and exciting atmosphere, while retaining and upgrading the original brand and reputation.

宽大的窗口让大自然得以进入空间内部。盆栽蔬菜和香草被悬挂在天花板上，摆放在架子上。特别是摆放在空间中央的植物架，它既供人观赏，又能被面包师所使用。空间通过家具分割，让各个角落都感觉十分舒适，整个餐厅感觉就像是祖母家的餐厅。在保持原有品牌和知名度的前提下，"旧时光"烘焙餐厅成功地营造出一种独特的氛围，令人惊喜。

Bakery MAISQUEPAN

梅斯科潘烘焙坊   Galicia, Spain
西班牙，加里西亚

Bakery in front of the busy street - breaking the image of traditional bakery

临街烘焙坊——打破传统烘焙坊的形象设定

Basic information 基本信息
Design/ 设计 : NAN Architects
Photography/ 摄影 : Iván Casal Nieto (www.ivancasalnieto.com)
Completion/ 完工时间 : 2014

Key materials 主要材料
All materials are locally created
所有材料均为本地生产

The project stems from the idea of the owners to create a multifunctional facility, which seeks to enhance their bakery business, incorporating the cuisine of Portuguese and Brasilian origins, offering suitable for a short stay tasting their products and coffee areas. Besides typical bread sale, the revenue would be very high.

The owners had some concerns about the design of the facility, and they did not want a conventional bakery and sought the establishment itself that gives them a good projection as brand image.

项目的出发点是业主想要打造一个能促进烘焙生意、兼具多重功能的场所：融合葡萄牙和巴西美食，为顾客提供能短暂休息、品尝美食和咖啡的区域。除了典型的面包销售额之外，总收入也会大幅增加。

业主有些担心空间的设计，因为他们不想要一个传统的烘焙坊，而是希望打造一个能够展示品牌形象的独特空间。

Both outside and inside the steel is used as a search for the material forward aseptic feeling.

Inside, it seeks to turn the space into a warm atmosphere, where people feel cosy and also look for items that evoke the tradition at the bakery, which is why the interior is lined with natural wood of spruce and some detail in decorative ceramic tiles.

The property has a kitchen and toilets of hiding in the back where the finishes are more conventional and industrial.

无论是室内还是室外,钢材的运用都体现了材料的美感,给人以干净的感觉。

室内试图营造出一种温暖的氛围,让人们感到舒适自在,同时也兼具一些传统烘焙的元素,包括天然云杉木、装饰瓷砖的细节等。

厨房和卫生间隐藏在空间的后部,它们的装饰更加传统而具有工业气息。

Lighting Design

Lighting also seeks a careful design. "We have sought shelter metaphorically evokes a large mass and this we have done with molds of the same material that mimic those of a loaf of bread; in the work area, these molds are an organic element where the 'mass' seems to fall to the ground. We have also opted for signage lighting and concealed lighting to enhance the wood panelling, and lighting in all the cases."

照明设计
照明同样需要精心的设计。"我们力求实现一种大空间的感觉,利用同种材料来模仿面包的纹理。在工作区,嵌灯呈现为有机的造型,坠坠欲滴。我们还选择了招牌照明和隐藏式照明来突出内部的木护墙板,并且为所有展示柜都提供了照明。"

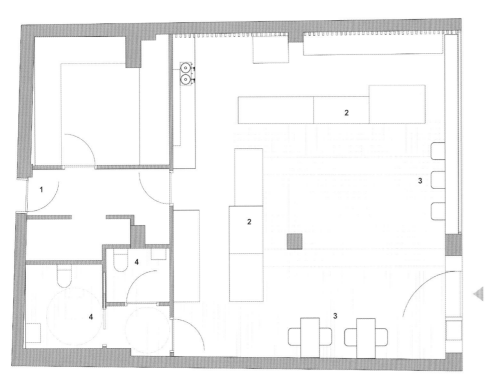

Floor Plan
1. Entrance
2. Counter
3. Seating area
4. Bathroom

平面图
1. 入口
2. 柜台
3. 就餐区
4. 卫生间

Bartkowscy Bakery

巴特科沃斯基烘焙坊　Toruń, Poland
波兰，托伦

Listen to the centenarian story of the family's craftsmanship
聆听家族工艺的百年故事

Basic information　基本信息
Design/ 设计 : mode:lina architekci architecture studio (Paweł Garus & Jerzy Woźniak)
Area/ 面积 : 60 m²
Photography/ 摄影 : Marcin Ratajczak (http://www.pfmarcinratajczak.fott.pl/)
Completion/ 完工时间 : 2014.9

Key materials　主要材料
Enamel tilling (floor); wooden plane (counter)
陶瓷地砖、木板台面

Bartkowscy is a family bakery of 18 stores in Toruń (Poland) with tradition that goes back to 1927.

In 2014 in cooperation with mode:lina architekci the owners created a new place with unique character on Aleja Solidarnosci in Toruń. The meeting with centenarian story of this family's craftsmanship was a huge inspiration for the new interior design. An aroma of fresh bread, huge ovens and old worn-out baking sheets made it clear to the architects that the fusion of tradition and modern design is the answer.

巴特科沃斯基是一家家族烘焙坊，在波兰的托伦市有 18 家分店，其历史可追溯到 1927 年。

2014 年，烘焙坊的经营者与 mode:lina architekci 事务所合作，打造了一个全新的分店。这家百年老店的故事对室内设计起到了重大的启发作用。面包的香气、巨大的烤箱和旧烤板充分体现了建筑师对传统与现代设计的融合。

Floor Plan
1. Entrance
2. Counter
3. Dining area
4. Kitchen

平面图
1. 入口
2. 柜台
3. 用餐区
4. 厨房

A complement of whole is white-black mosaic floor that is the reminiscent of old stores. Traditional atmosphere is also omnipresent thanks to oven-shaped furniture and natural wood.

黑白相间的瓷砖地面令人想起旧日的商店。烤箱型家具和天然木材的运用让传统氛围无处不在。

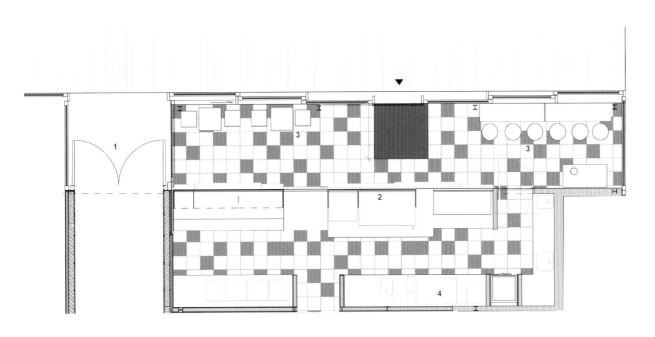

Dark patinated baking sheets received a second life as a wall décor and lamp shades.

The counter and lighting are finished by meticulously selected wooden planks that may refer to traditional oven fuel.

斑驳的黑色烤盘在墙面装饰和灯罩中获得了新生。

柜台和照明设施的设计精心搭配了木板，令人想起传统烤箱的燃料。

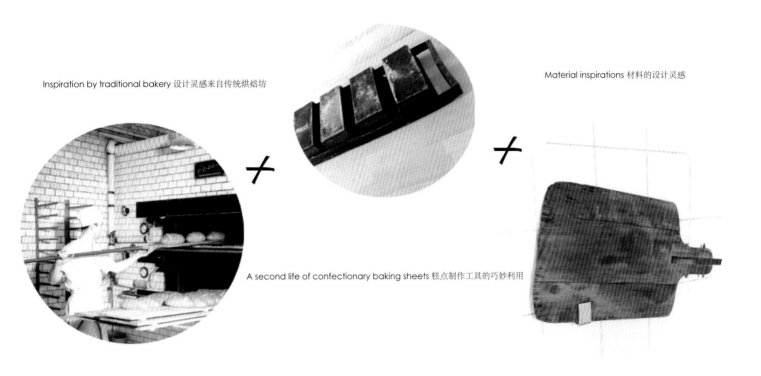

Inspiration by traditional bakery 设计灵感来自传统烘焙坊

Material inspirations 材料的设计灵感

A second life of confectionary baking sheets 糕点制作工具的巧妙利用

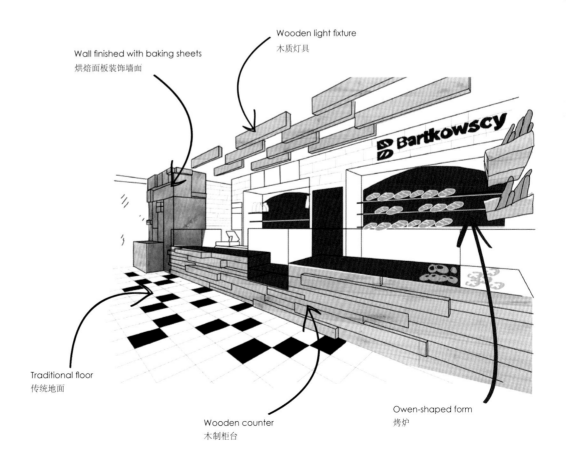

Wall finished with baking sheets
烘焙面板装饰墙面

Wooden light fixture
木质灯具

Traditional floor
传统地面

Wooden counter
木制柜台

Owen-shaped form
烤炉

Successful Bakery Design II 139

BREAD, ESPRESSO&

BREAD, ESPRESSO& 面包咖啡店
Taipei, Taiwan, China
中国，台湾，台北

Real power in a quiet alley
在小巷弄里展现真实力量

Basic information 基本信息
Design/设计：Aki Hamada Architects + Kentaro Fujimoto
Area/面积：111.6m²
Completion/完工时间：2015

Key materials 主要材料
Red cedar, cement board (façade)
红松木、水泥板（立面）

BREAD, ESPRESSO& has been ranked as one of the top 10 delicious toast by Premium, a Japanese magazine. This is the brand's first overseas brand. When asked why locate it in Taiwan, Vera, the store manager said: "Not only because Taiwan is near Japan, but also in consideration of culture and dietary habit."

This small store takes a Japanese baker as well as the spirits of Omotesando, Japan to Taiwan. Hidden in this alley, BREAD, ESPRESSO& expresses its real power.

BREAD, ESPRESSO& 面包咖啡店入选了《Premium》十大好吃吐司。如今在台湾开了首家海外分店，为何把分店开在台湾呢？店长Vera说："不只因为台湾与日本地理位置近有关，文化与饮食习惯更是重点之一。"

这家小店带来了日本师傅棺原秀，也将东京表参道的精神输入台湾，正是隐身直横交错的小巷弄里，更显BREAD, ESPRESSO& 朴实无华的真实力量。

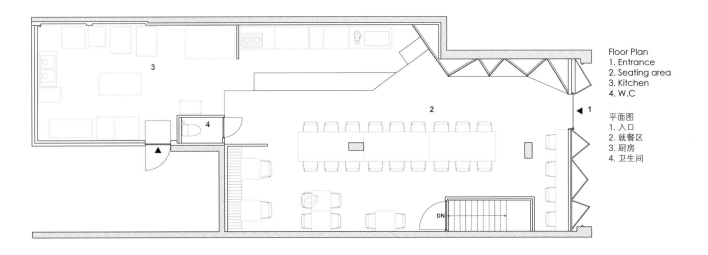

Floor Plan
1. Entrance
2. Seating area
3. Kitchen
4. W.C

平面图
1. 入口
2. 就餐区
3. 厨房
4. 卫生间

As for the interior, wooden tiles of the same size as the exterior tiles are arranged, and triangular walls are finished with wooden tiles and mirrors to have two distinct sides like the outer walls, so that you will have totally different impressions when entering and leaving the shop. It was aimed to be a dynamic shop in terms of both the exterior and interior which continuously changes its impression depending on where you see it from.

在室内设计中，尺寸大小一致的木砖得到了大量的应用。它们与镜子构成了与室外空间相似的三角形墙壁，让进店和出店的体验变得截然不同。无论是室内设计还是室外设计，店铺的目标是通过不断变化的形象给人以充满活力的感觉。

BINARIO 11

拜纳里奥 11 号　Milan, Italy
意大利，米兰

The romantic trip in the train station
火车站里的浪漫旅程

Basic information 基本信息

Design/ 设计：Andrea Langhi Design
Design team/ 设计团队：Andrea Langhi, Samanta Volpi, Santo Scibetta, Samuele Bernasconi
Area/ 面积：150 m²
Photography/ 摄影：DANIELE DOMENICALI (www.danieledomenicali.com)
Completion/ 完工时间：2013

Key materials 主要材料

Brick (Wall)
砖（墙面）

Binario 11 is home to the main train station with ten tracks in Milan.

The counter is constructed over old suitcases, everything to celebrate the journey and its most romantic and evocathive vehicle: the train. The interior design refers to a vaguely exotic atmosphere with a mixture of styles and different influences. Memories almost completely forgotten but which worth remembering in this space behind the Cadorna railway station, which has ten tracks. This place is called Binary 11, the track that does not exist: that of fantasy.

拜纳里奥 11 号烘焙坊坐落在米兰的 10 轨火车站。

烘焙坊的柜台下方堆满了旧皮箱，这些都与旅行以及最浪漫的交通工具——火车相关。室内设计呈现出暧昧的异国氛围，融合了各种不同的风格。人们在这里几乎遗忘了一切，唯一记得的是这里在卡多尔纳火车站后方。这里是拜纳里奥 11 号，这里没有铁轨，只是一场迷梦。

The venue is placed in an old mansion and it is divided into two different areas: the bakery and the dining room.

烘焙坊坐落在一栋旧公寓里,分为两个区域:烘焙区和就餐区。

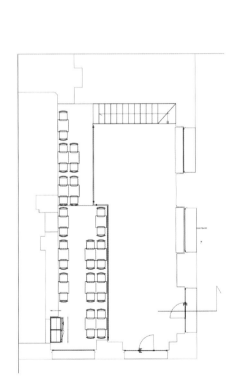

The first one features a space adorned with hanging Renaissance reproductions from Agnolo Bronzino where the floor has the best Italian Intarsia method. A black back wall with poems from Giovanni Pascoli explaining the process of making bread, combines with the original structural bricks and industrial lighting to create a whole new way of displaying food, celebrating the goodness of Italian bakery. Everything to remember the passion for the beautiful and good things that make every day a special day.

烘焙区悬挂着保罗·布隆奇诺文艺复兴时期作品的复制品，地面采用了最好的意大利木造镶嵌工艺。黑色的背景墙上书写着乔瓦尼帕斯科里的诗，描绘了面包制作的过程；背景墙与原始的结构砌砖和富有工业气息的灯具共同营造出展示美食的新方式，体现了正宗的意大利烘焙。请记住，美好的事物让每一天都变得很特别。

The second room is a bit more romantic, trying to bring back the nostalgia from the Orient Express era. The floor here is a black and white chessboard and there is a clever mixture of styles becoming even a bit exotic. The bar has an old distressed look with a pewter top. There is a mezzanine level with tables and chairs overlooking the ground floor where the guest can appreciate the scale of the big round lanterns illuminating the space.

就餐区更加浪漫,试图带领我们回到东方快车的时代。地面由黑白双色棋格拼接而成,各种风格的巧妙融合甚至呈现出一点异域风情。吧台采用青灰色台面,有些破败的感觉。客人可以坐在中层楼的座椅上俯瞰楼下,欣赏巨大的圆形灯笼。

DESIGNER'S VISION
设计构想

Colour and forms serve as 'input' to attract customers walking in
以色彩与形态来吸引顾客走进来

Before designing this project we´ve visited and analysed other similar spaces trying to find some errors that could be corrected. We found out that a basic error being committed was that most of these services only had one type of space. This design attitude ignored the variation of mood one fells during the day, or even if he walks there alone or with friends, needs a place to read a book or just wants to socialise. So, to bridge this flaw, we created three different environments so that the costumer can select the space that fits better to his or her mood, rather than have to adapt himself or herself to an imposing environment. This way we provide a more emphatic place and consequently amplify three times the commercial potential.

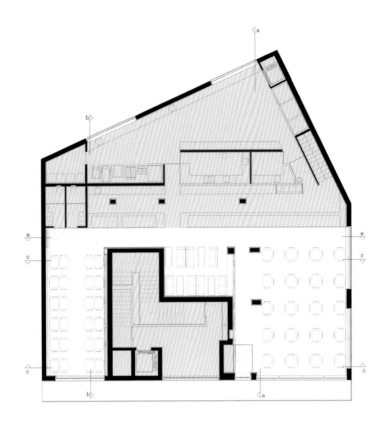

在设计这个项目之前，我们走访了其他类似的空间，期待能找出一些错误，以便改正。我们发现，一个基本的错误就是，不同的服务都放在同一类型的空间。这种设计忽视了我们一天之中情绪的变化，也忽视了"独自一人"和"与朋友相伴"两种情况的区别——前者你可能需要一个安静的地方读一本书，后者更需要交流的空间。因此，为了避免这种错误，我们打造了三种不同的环境，顾客可以选择更适合自己情绪的环境，而不是根据既定的环境来调整自己。这样，我们就提供了一种多样化的空间，进而将店铺的商业潜能扩大了三倍。

But a customer is not one till he gets in. How could we get him inside?

In a metropolitan style of life, everyday people deal with millions of inputs, like Billboards, Signs, People, Cars...The way the brain deals with this excessive information is to send most of it to the unconscious mind, releasing the conscious from the excessive information. As one moves through the city the brain captures the information around and gathers all the similar inputs creating a mental 'scenario' that, based on predictability is perceived by the unconscious mind, releasing the conscious to all variable inputs that he experiences outside that scenario. This is a surviving system that we inherited from the savanna era, so that if for example, a predator moved between the trees, without having to consciously capture every bit of information around, one could perceive the movement and react to protect its own life.

Joining to this line of thought the known fact that 70% of those inputs are visual, and that humans as many animals have an attraction to light, we knew that we had to create an input that could distinguish itself from the rest of the city scenario in such a way that it could activate the conscious perception, guaranteeing that people would notice and fell attracted to it. For that we´ve used light as the main attraction.

We've studied the approximation of the observer to the space and realised that the most visually relevant plan from the exterior was the ceiling and so, we focused on that.

In our studies we also realised that the use of direct light tends to heat up the space and create shadowed corners turning space into uninviting places and that, in an auditory approach, the excessive noise mainly resulted from the reverberating sound was not being properly solved.

So, to solve these problems we knew we had to break the sound waves and refract the light. And so we did; by creating a second ceiling that results from the repetition of wooden stripes, we found a system that could solve the two problems in a row.

但是，没走进店内，他就还不是顾客。那么，我们如何让他进来呢？

在现代都市生活中，人们每天都要处理大量信息，比如广告牌、标识、人、车……大脑处理过量信息的方式，就是将大部分信息放在潜意识中，让我们的意识不受过量信息的干扰。走在城市中，你的大脑不断接收周围的信息，并将所有相似的信息组成"脑内情境"，由潜意识来感知，而我们的意识则去处理那些"情境"之外的差异化的信息。这是我们从热带草原时代继承下来的生存方式，这样，如果捕食者在树林中移动，被捕食者就不必有意识地去接收周围的所有信息，而只关注捕食者的行动并做出反应，即可保护自己的生命。

顺着这个思路，我们还知道，70%的外界信息是视觉信息，而人类，跟许多动物一样，会受到光的吸引。因此，我们知道，我们必须创造出与周围城市环境的情境有所差异的东西，能让人们有意识地去接收，确保人们能注意到，并为之所吸引。为此，我们使用了灯光作为主要的吸引手段。

我们研究了观察者到空间的距离，发现从外面来看，最吸引视线的就是天花板。于是，我们将重点放在天花板。

通过我们的研究，我们还发现，直射光会让空间过热，还会产生"阴影角落"，使环境变得让人却步。从听觉的角度上说，光反射产生的过量噪声也无法恰当解决。

所以，为解决这些问题，我们必须打破声波，改变光的折射。我们的方法是创造第二层天花板，利用木板条的重复叠加，同时解决上述两个问题。

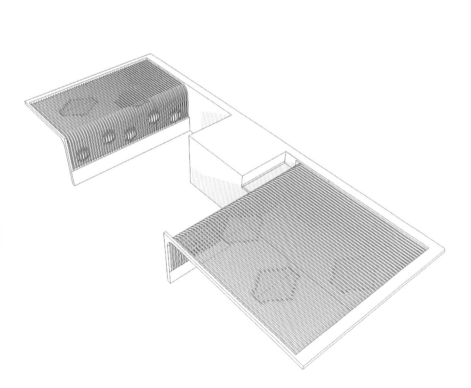

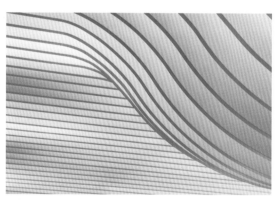

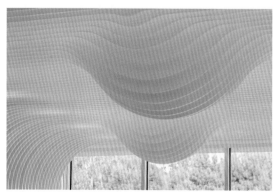

How do the colour and form make you happy?

In our research we´ve found studies that prove that the presence of colour and forms that are food alike actually makes people hungrier. So to get that input on the users, we´ve picked the twenty most wanted products of the bakery and based on a pattern of global identification we found a middle tone and applied it on the walls.

On the formal approach, we made the ceiling 'melt' in some points to make it look like a cake topping.

We also proposed a new logo to the client, and designed the space partially based on it. The wooden stripes descend through two of the walls creating an effect that dialogues directly with the consumer. When one moves through space he realises that some hidden forms start to appear on the walls. Those forms are an abstraction of the proposed logo. The intention is to unconsciously reinforce the image of the firm in one's mind.

We like to think our interventions as positive manipulation of the human brain. As such we focus on giving positive inputs to all the five senses (when possible) so that we can alter one's homeostatic level, and as a result make people feel happier.

Title/ 名称：BAKERY
Location/ 地点：Gondomar, Porto
Design/ 设计：Paulo Merlini arquitectura
Photography/ 摄影：João Morgado

色彩和形态如何令人愉快？

通过我们的研究发现，像食物一样的色彩和形态的存在能让人感到更加饥饿。所以，为了让顾客产生这种感觉，我们选择了这家烘焙坊20种最受欢迎的产品，又根据国际认知模式找到了一种中间色调，应用在墙面上。

在形态上，我们把天花板在某些点上做成"融化"状，看起来就像蛋糕上面的样子。

另外，我们还给客户设计了新的LOGO，一部分空间的设计也是以这个新LOGO为出发点的。木板条绵延到两面墙上，让人产生与空间直接对话的感觉。人在空间中走动的时候会觉得某些隐藏的形态开始在墙面上显现出来。这些形态就是新LOGO的一种抽象表现形式。设计目标是通过这种方法潜意识地强化这家公司在顾客心中的形象。

我们认为我们的设计是对人类大脑的一种积极操控。我们尽可能对顾客所有的五种感官产生积极的影响，以此来改变人的自我平衡状态，最终使人感觉更愉快。

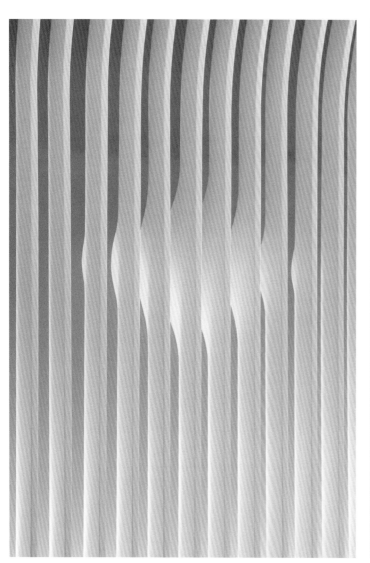 

Chapter 3 Display Design
第三章 陈列设计

3.1 Display Furniture and Methods
3.1 陈列家具与方式

3.2 Lighting for Display
3.2 陈列照明

Case Study
案例解析

Bread Table
日本岐阜面包桌烘焙坊

Pâtisserie À La Folie
疯狂点心店

Les Bébés Cafe & Bar
贝贝西点咖啡

Boulanger Kaiti
日本福冈卡提面包店

midi a midi
迷迪面包店

T by Luxbite
T 甜点屋

Style Bakery
风尚面包店

So Milaky
索·米拉其面包店

Serrajòrdia
塞拉约迪亚面包店

Chocolateria Brescó
布雷西科巧克力烘焙坊

IL LAGO Bakery & Wine Shop in MVL Hotel Kintex
韩国国际会展中心 MVL 酒店湖泊烘焙坊与酒廊

Gail's Bakery, Chelsea
英国伦敦切尔西盖尔面包店

Chapter 3
Display Design

第三章
陈列设计

3.1

3.2

When the baked goods are ready to sell, it's essential that they earn your customers' attention. As the saying goes, the product needs to look 'good enough to eat'. Presenting an attractive display is one way to garner that visual appeal from customers and passersby. Learn some practical and useful tips for displaying bakery items is essential.

3.1 Display Furniture and Methods

Choose suitable display cases

Bakeries typically use bakery display cases to show off their goods to the public. This is one of the first things to purchase when planning on selling bakery items retail to customers. A dry bakery display case or refrigerated bakery display case can be chosen, depending on the product to display. (See figure 3.1, figure 3.2)

Dry bakery display cases are meant for items that require no refrigeration. These are ideal choices for breads and pastries. Often, pastries or desserts with fresh dairy components, such as whipped cream, custard or fruit topping require a refrigerated holding place. These are higher-end products that have high-quality yet easily perishable ingredients. Baked goods that include cheese or meat must also be refrigerated, such as ham and cheese croissants. Both types of display cases provide a secure method of high-volume merchandising for effective display. Bakery display cases are a must for keeping baked goods in a secure, clean environment until they are be purchased and consumed.

Try a display stand

Although many bakery items are perfect when displayed in a bakery display case, others are better suited to tiered food display stands, cake stands or display pedestals. These types of display pieces are designed to elevate food and make it not only more noticeable, but also more appealing to guests. Choose from clear glass, plastic or metal display stands and pedestals.

烘焙食品制作完成后，要让这些食品吸引顾客的注意。卖相要好，起码要"看上去很好吃"。美观的陈列是增加食品对顾客和路人吸引力的一种重要方式。以下介绍一些有关烘焙食品陈列的实用技巧。

3.1 陈列家具与方式

选择适合的陈列柜

烘焙坊一般使用烘焙陈列柜来展示食品。烘焙食品零售设计中这是首要考虑购买的东西。普通烘焙陈列柜或者烘焙陈列冷柜都可以选择，取决于陈列食品的种类。（见图3.1、图3.2）

普通烘焙陈列柜主要用于无需冷藏的烘焙食品展示，是面包和糕点的理想选择。通常，带有新鲜奶制品的糕点或甜品（比如生奶油、奶油冻或者水果）需要冷藏陈列。这些属于高端商品，配料考究，易变质。含有奶酪或肉的烘焙食品也必须冷藏，比如火腿面包和奶酪牛角面包。这两种陈列柜都能满足大容量、安全陈列的需求。烘焙陈列柜是保证在购买和食用发生之前烘焙食品在安全、清洁的环境中保存的必要手段。

尝试使用展示台

许多烘焙食品摆放在烘焙陈列柜里非常完美，但是也有一些更适合分层展示台、蛋糕架或者独立式展台。这些类型的展示工具，设计意图是将食品摆在高处，不仅更显眼，而且让顾客看着更有吸引力。可以从透明玻璃、塑料或金属展台中选择合适的来应用。

让陈列更丰富

3.3 3.4 3.5

Create a scene of abundance

Just about any type of baked treat, whether loaves of bread or chocolate chip cookies, look better when there are a lot of them all displayed together. When the bakery case or display stand only has a meager selection of items available, they become less appealing to the customer. Make sure the bakery display is fully stocked, or create the illusion of showing more than what have with mirrored walls inside the display case. Of course, smaller display schemes can also be used, such as smaller baskets overflowing with bread, or smaller display stands stacked and layered with brownies. (See figure 3.3)

Consider using colour

Think about what colours the baked goods are, and what colours might complement them. For instance, many baked items are brown, cream-coloured, tan or golden in colour. Although the items will probably look appetising anyway, choosing colours to complement baked goods can be a smart move. Colour theory suggests that warmer colours such as reds, oranges, yellows and browns can help stimulate the appetite. Backing the baked goods with coloured cloth, coloured display stands or even colourful décor like faux flowers or beads can act to enhance the natural appeal of bakery items. Many establishments stick to a certain colour scheme. In this case, using these colours to display baked goods makes sense as well. (See figure 3.4, figure 3.5)

Keep everything clean

Keeping things looking neat, clean and fresh is an essential aspect of displaying bakery items. Cookies need to look soft and fresh. There should be no visible grease stains on tissue paper, nor excessive crumbling on display platters. Use fresh wax paper or tissue paper below pastries and cookies so that everything looks as fresh as possible. As far as bread goes, you are likely only displaying

任何种类的烘焙食品，不管是长面包还是巧克力曲奇，都是较多数量摆放在一起的时候看起来效果更好。如果烘焙陈列柜或者展台上可供选择的食品种类有限，那么在顾客看来就没有那么诱人。要确保陈列柜是摆满的状态，或者可以在陈列柜内部安装镜面，营造出陈列品比实际更丰富的错觉。当然，还可以增加比较随意的陈列方法，比如用小篮子，里面盛放面包，或者小展台，上面分层叠放布朗尼蛋糕。（见图3.3）

使用色彩

想想烘焙食品都是什么颜色的，使用哪些颜色能与之协调。比如说，很多烘焙食品是棕色、奶油色、黄褐色或者金色的。这些食品本身看上去已经够让人产生食欲了，再选择与之和谐的色彩来搭配，效果会更好。根据色彩理论，暖色（比如红色、橘色、黄色和棕色）能够刺激食欲。以某种色彩的布料、展台或者甚至是色彩鲜艳的装饰品（如假花或珠子）作为烘焙食品的陈列背景，能够强化这些食品的吸引力。很多烘焙坊会专门使用某种固定的配色。这样的话，烘焙食品的陈列也可以使用其中的颜色。（见图3.4、图3.5）

保持干净整洁

烘焙食品的陈列，一切保持干净、整洁、新鲜的状态，这点很重要。曲奇要看上去柔软、新鲜。包装纸上不应有肉眼可见的油迹，盛放食品的托盘上也不应有太多碎渣。糕点或曲奇下面可以使用干净的蜡纸或棉纸，整体看起来会更加整洁清新。面包当然是尽可能陈列新鲜的，但如果不是新鲜的，至少也不应该让人看到表

Chapter 3
Display Design

第三章

陈列设计

3.6

3.7

fresh product, but if not, there should be no sign of mold on the crust. Additionally, be sure that all glass surfaces on the bakery display cases are clean and free of fingerprints. Keep an eye out for flies and other invading pests.

3.2 Lighting for Display

The purpose of lighting an object is to make it more appealing. Usually bakery display cases have some sort of lighting built in to accentuate the goods by enhancing colours and overall appearance. Lighting can add life to bakery display, as long as you do not over- or under-light the most important areas. Customers should be able to easily see textures and colours without straining. Be careful of using hanging light fixtures that may spotlight the item and wash it out under too much light. These types of lights may also produce excessive radiant heat, warming the item and potentially drying it out or disturbing its consistency. Instead, go with ambient light, such as from a fluorescent light in a bakery case, or make use of daylight through windows whenever available. (See figure 3.6, figure 3.7)

Keep in mind of colour

Light has colour. Sunlight and LED lights have a blue cast. Lamplight has a yellow cast. Fluorescent lights have all sorts of colour casts.

The most important thing to remember is consistency. Try to keep case lighting even by using the same type of bulb throughout all the cases. This presents a clean look for all the products displayed. Inconsistent case lighting breaks down an item's ability to reach its full visual potential.

If it is an upscale bakery and the cases are all illuminated with slightly different colours, LED lights are a growing and promising technology. The daylight-balanced bulbs and fixtures illuminate at a blueish colour that is similar to that of the sun coming through windows.

面有霉菌。另外，确保陈列柜的玻璃干净清洁，上面没有手指印。注意店内是否有苍蝇或其他飞虫。

3.2 陈列照明

给某个东西增加照明，目的是让它更具吸引力。通常，烘焙陈列柜里会有某种照明装置，突出里面陈列的食品，使其颜色和整体观感更加诱人。照明能让烘焙食品的陈列更有生机，只要你不要给最重要的区域过多或过少的光照。顾客应该不必眯眼细看就能看出烘焙食品的质感和颜色。使用悬垂的灯具要小心，悬垂灯下方的食品会自然而然地成为焦点。过多的光照会产生热辐射问题，等于给下面的食品加热，烘干其中水分。使用背景光就会好得多，比如在陈列柜里安装荧光灯，或者尽可能利用窗口的日光。（见图3.6、图3.7）

注意色彩

光是有颜色的。日光和LED光是蓝色的，普通照明灯是黄色光，荧光灯的光可以是各种颜色。

需要注意的最重要的一点是一致性。所有陈列柜要使用相同类型的灯泡，确保陈列柜里光照的均匀、一致。这样，陈列的所有食品看上去会更整洁。如果陈列柜的照明不一致，某一陈列品会无法充分展现其视觉魅力的潜能。

如果是高档烘焙坊，陈列柜的照明灯光颜色略有不同，那么可以使用LED灯。LED照明是一种快速发展的新兴技术。LED灯光很像日光，偏蓝色，看起来就像从窗口洒入室内的自然光。

If there are nice windows already, then adding some daylight-balanced LED fixtures to cases is a great start to colour consistency. Stay away from fluorescents.

Thinking about the impact of lighting on freshness & quality

Customers may notice the smell wafting from the bakery but if the bread looks yellowy and is rock hard, chances are the customer will think twice about placing that item in her basket. The type of lighting to select in the bakery plays a very crucial role in the freshness and quality of the product.

All the delicate cream fillings and icings that decorate cakes and pastries deteriorate and lose their vibrant colours under regular lighting. Colour distortion of icing and cakes not only represents a significant dollar loss for the retailer but also indicates a more serious problem with the display lighting and the impact it is having on the products on display.

Regular light also affects the fragrance and flavour of fresh baked products. All the time and energy that went into creating such delectable treats is wasted when they are displayed under inferior quality lighting.

It is very important to select appropriate lighting fixtures for display in a bakery, such as specialty-merchandising lamps or food-safe lighting, which reduces UV radiation and protects the shelf life of all the bakery goods.

如果开窗条件良好，那么可以在陈列柜中增加一些LED灯，保证陈列柜照明的一致性。避免使用荧光灯。

照明对食品质量的影响

顾客可能会被从烘焙坊中飘出的香气所吸引，但是，如果面包看起来是浅淡的黄色，如岩石一般坚硬，那他们可能要再想一想要不要把这样的面包放到自己的购物篮里了。烘焙坊中照明灯种类的选择很关键，会影响到顾客对食品质量的感知。

所有装饰蛋糕和甜点的精致奶油花样和糖衣酥皮在普通的照明下都会失去其诱人的色泽，看起来平凡无奇。糕点的色彩失真不仅意味着店家生意的损失，而且提醒了我们关于陈列照明的一个更严重的问题，昭示了照明对陈列食品的影响。

普通照明也会影响到新鲜烘焙食品的香气和味道。如果陈列在普通的室内照明条件下，所有花在制作这些精美食物上的时间和精力就都白费了。

选择合适的烘焙陈列照明装置也非常重要，比如特色推销照明灯或者食品安全照明灯，后者能减少紫外线辐射，延长烘焙食品在陈列架上保存的时间。

Figure 3.1, figure 3.2: The distinctive display methods can enhance the experience of customers in bakeries
图 3.1，图 3.2：独特的陈列方式能够增强顾客在店内的体验
Figure 3.3: The display panel on the wall looks simple, but creates an effective method as well as a neat looking
图 3.3：墙壁上的陈列台看似简约，但却是一种有效的方式，同时营造出整洁之感
Figure 3.4, figure 3.5: The warm colour of the display stand makes the baked products more delicious
图 3.4、图 3.5：暖色调的陈列架使得烘焙食品看起来更加美味
Figure 3.6, figure 3.7: The lighting fixtures play an important role in baked products display
图 3.6、图 3.7：照明对于烘焙食品陈列起着至关重要的作用

Bread Table

日本岐阜面包桌烘焙坊 Gifu, Japan
日本，岐阜

Less is more – less materials, more beauty
少即是多——更少的材料，更多的美

Basic information 基本信息

Design/ 设计 : Keiichi Kiriyama / Airhouse Design Office
Area/ 面积 : 46.37 m²
Photography/ 摄影 : Toshiyuki Yano

Key materials 主要材料

Plywood, translucent polycarbonate, steel frame
胶合板、半透明聚碳酸酯板、钢架

The second bakery to feature on Dezeen is designed by Japanese studio Airhouse Design Office and features a tree growing out of its curved timber counter.

Located in the central Japanese prefecture of Gifu, Bread Table is a small bakery with a shop space and kitchen divided by a structural plywood display counter.

第二个登上《Dezeen》设计杂志的烘焙坊，由日本"充气屋设计工作室"（Airhouse Design Office）设计完成。设计的特色是从木板柜台上"生长"出来的一棵树。

店铺位于日本本州岛中部的岐阜县，面积很小，由零售空间和厨房两部分组成，两者用胶合板展示台分隔开来。

Cane baskets piled with loaves of bread and wire racks of pastries are stacked at intervals along the counter, while translucent polycarbonate corrugated sheets line the front and give off a pink glow when the room is lit up in the evening.

The same corrugated sheets have also been used to line a wall and the interior of the door, which features a chunky wooden handle.

The kitchen and selling space are considered to have equal weight, and a large table-like platform is created between them. "The plywood counter can be used for a variety of purposes such as a display space, checkout counter or a working space to cut bread and knead dough," said architect Keiichi Kiriyama.

Structural plywood was used for the top of the table, and translucent polycarbonate corrugated sheets are placed on the sides of the platform. By designing the sides as translucent, it is intended to emphasise the presence of the table while blurring boundaries.

For this shop with a small-sized staff, this design enables the owner to always have knowledge of the shop situation and allows different actions depending on the amount of bread produced. As a result, this table creates an open atmosphere, fosters communication between the customers and bakers, dynamically displays the process from the time the bread is baked to the moment it is sold, and creates a rich retail space that is fun for working as well as shopping.

面包盛放在藤条编织的篮子里，糕点摆放在柜台上的货架上，半透明的聚碳酸酯波纹板装饰着柜台正面，傍晚开灯后，营造出淡粉色的、散发出柔光的空间环境。

还有一面墙上以及门的内侧也采用了相同的波纹板，门上使用了敦实的木质把手。

设计师认为厨房和零售空间同等重要，并在二者之间设计了一个大型柜台。设计师桐山庆一表示："这个胶合板柜台可以有很多用途，比如用作展示、登记或者切面包、揉面的工作台。"

柜台顶面采用了胶合板，侧面则使用半透明聚碳酸酯波纹板。侧面设计成半透明，意图是凸显柜台的存在感，同时模糊厨房和零售空间二者之间的界线。

这是一家小店，店员也不多，长柜台的设计确保了店主随时能看到店内的所有情况，并且能根据制作面包的数量来改变销售的方式。同时，这个柜台也营造出室内通透的空间氛围，促进了烘焙师傅和顾客之间的交流，让面包从烘焙到销售的整个过程清晰可见，大大丰富了零售空间的内容，在这里工作或购物都成为有趣的过程。

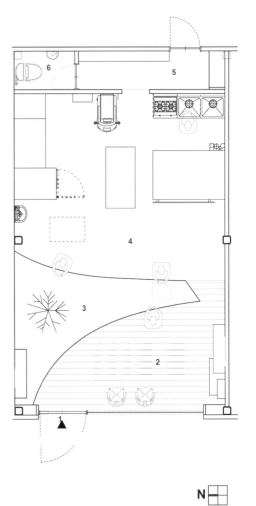

Floor Plan
1. Entrance
2. Shop space
3. Display table
4. Kitchen space
5. Storage
6. W.C.

平面图
1. 入口
2. 零售空间
3. 展示台
4. 厨房
5. 储藏间
6. 洗手间

Pâtisserie À La Folie

疯狂点心店 Montreal, Canada
加拿大，蒙特利尔

A pastry defies all the typecasts of traditional pastry shop
颠覆传统的点心店

Basic information 基本信息

Design/ 设计 : Atelier Moderno (ateliermoderno.com) + Anne Sophie Goneau (asgoneaudesign.com)
Area/ 面积 : 60.4 m² (650 sq.ft)
Photography/ 摄影 : Stéphane Groleau
Completion/ 完工时间 : 2014

Key materials 主要材料

Wood cladding, steel display
木包板、钢展架

The design of the new address of the Pâtisserie À La Folie aims to defy all typecasts of the conventional pastry shop by giving full prominence to the superior presentation and quality of its unique products.

The execution of the design begins with the storefront. The shop's 650 sq.ft. ground floor location on the colourful street of Mont-Royal impelled the creation of a space that is distinguished by its neutrality, acting as a monochromatic backdrop to the vibrancy of both the street and the pastries within.

疯狂点心店的设计目标是颠覆传统点心店的类型，充分突出其独特产品的特色和品质。

设计从店面开始，整个店铺总面积60.4平方米，坐落在色彩斑斓的皇家山大街上。店面设计简洁中性，单一的色调与整条大街和店内点心的多姿多彩形成了鲜明的对比。

All pastries are made in-house, with the production floor set in the rear of the space. It is separated from the shop area by a mid-height wall that encourages the sounds and smells of the kitchen to tease the senses of the patrons beyond. A theme of assembly line production is demonstrated by a band of hemlock slats that extend above the dividing wall and span the full depth of shop space, recalling the aesthetic of the wooden pallet or the conveyor belt.

点心店里的所有糕点都现场制作。制作区就设在后方,通过一面中等高度的墙壁隔开,厨房的香气和制作的声音不断地飘出来挑逗顾客的感官。背景的铁杉木板条呈现为传送带的样式,一直向上延伸到隔断墙上,切入整个店面空间,令人想起木托盘或传送带的奇特美感。

Floor Plan
1. Entrance
2. Display counter
3. Kitchen/workshop
4. W.C.

平面图
1. 入口
2. 展示柜台
3. 厨房 / 制作间
4. 洗手间

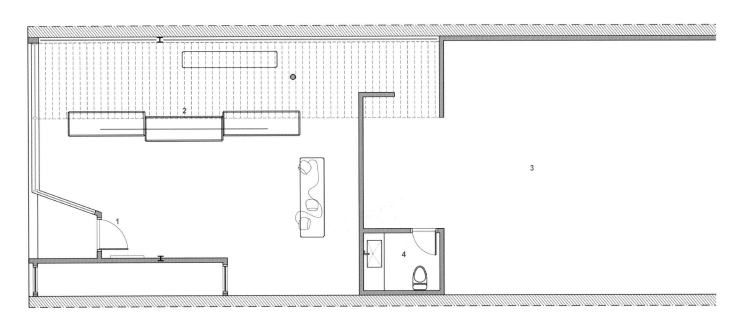

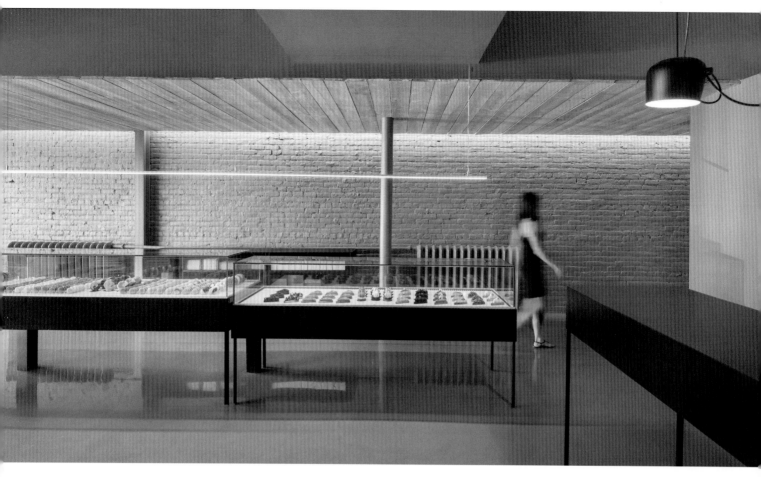

The refrigerated display units sit below the wood cladding, while thin and delicate LED suspension defines this exhibit area with its focused and dramatic illumination. The black steel display units were custom designed and fabricated to stand independently; the compressors for each unit are positioned in the basement below, allowing the displays to stand on dainty steel legs, in continuity with the adjacent storage and cash counter units. The cash counter itself is accentuated by the chaotic lighting of the Flos pendant lamp.

冷藏展示柜设置在木包板的下方,头顶上纤细精致的 LED 灯用聚焦的灯光突出了展示区。黑钢冷藏展示柜是特别定制的,能独立站立。冷藏柜的压缩机位于下方的地下室里,使得地面上只有简单的四根桌腿支撑,与相邻的储藏柜以及收银台实现了统一的设计。Flos 吊灯的散射灯光突出了收银台的简洁质感。

Les Bébés Cafe & Bar

贝贝西点咖啡 Taipei, Taiwan, China
中国，台湾，台北

Experience the spatial concept of 'folding'
体验"折"的空间概念

Basic information 基本信息

Design/ 设计：Johnny Chiu (lead), Nora Wang, Sunny Sun
Area/ 面积：150 m²
Photography/ 摄影：Kyle You
Completion/ 完工时间：2013.3

Key materials 主要材料

White/black laminate flooring; ceiling: painting colour black, gray and white; wall: Mexion board; counter top: ariston marble
地面：黑白色复合地板；天花板：黑、白、灰色涂料；墙面：Mexion 板材；台面：Ariston 大理石

"Upon the success of the first Les Bébés Cupcakery store using the simple gesture of folding in creating the store space, the 'folding' action became very interesting to us and we wonder how we can recreate this spatial concept without again to fold out space, but still be able to continue this concept in our second store." – Words from the designer.

延续"贝贝西点一店"的成功，由"折"包装盒发展出的纯净空间，成为展演杯子蛋糕的最佳舞台。面对二店"贝贝西点咖啡"，我们思考如何延续"折"的概念及品牌精神，在不同的空间，做不同的动作。——设计师的话。

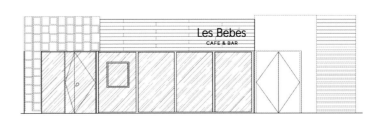

For customers, this sense of delight, surprise and freshness of transformation from the entrance into its take-out area, into its cafe area is an ever-changing scope that once again allows architecture and Les Bébés products to sync and complement each other.

对顾客而言，从门外经过外卖区再到内用区，伴随感官上的愉快，惊喜及新鲜感，一如青田店，简约的配色及欢愉的气氛，让杯子蛋糕及餐点成为主角。建筑空间、西点及其包装再次合为成功的西点品牌。

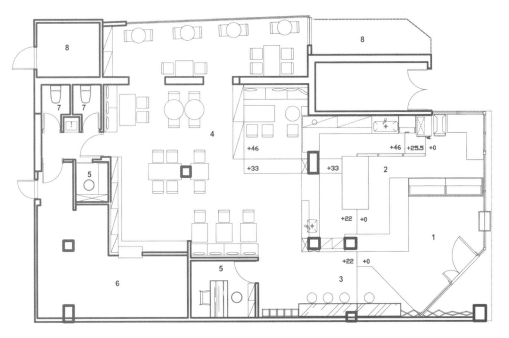

Floor Plan
1. Take-out area
2. Take-out kitchen
3. Bar
4. Dining area
5. Office
6. Kitchen
7. Toilet
8. Machine room

平面图
1. 外卖区
2. 外卖厨房
3. 吧台
4. 用餐区
5. 办公室
6. 厨房
7. 洗手间
8. 设备间

Section
剖面图

Bar 吧台
Folded box wall and ceiling 折叠式墙面与天花
Dining area 用餐区
Kitchen 厨房

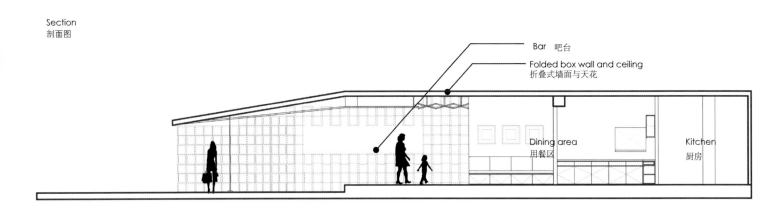

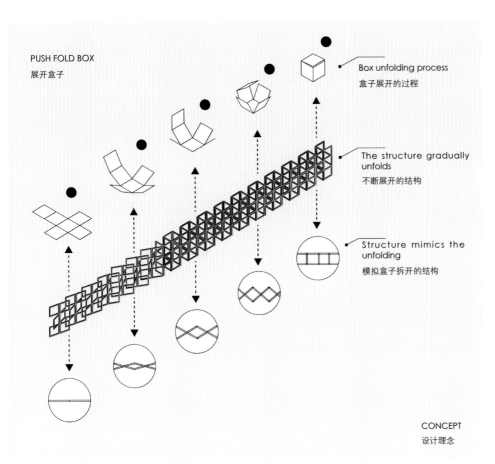

PUSH FOLD BOX
展开盒子

Box unfolding process
盒子展开的过程

The structure gradually unfolds
不断展开的结构

Structure mimics the unfolding
模拟盒子拆开的结构

CONCEPT
设计理念

The designers relook at the packaging box, and wonder what if they record the idea of folding, so it's about the 'action', rather than the 'result'.

They took the box and simplify it down to its structural elements. With four different folding steps, 0, 30, 47 and 90 degree angle, they folded the box from its original state to a 3D object. The designers created a series of these boxes in these four angles so the transition from the exterior of the shop to the interior is a journey of discovery and excitement.

设计师重新回到包装盒，借由不断地拆解及组合的"动作"衍生出一连串的纪录，这次并非单纯的折出一个空间，而是要在这空间中记录包装的过程，同时呼应了每位顾客来店里的一段体验。

他们拿掉盒子的面，让它只剩框架，借由四个角度折迭的步骤，0度、30度、47度和90度，从平面到立体，创造出一连串角度变化的过程，从外观立面到内部立面及天花，像是一段从外到内探索美食的小小旅程。

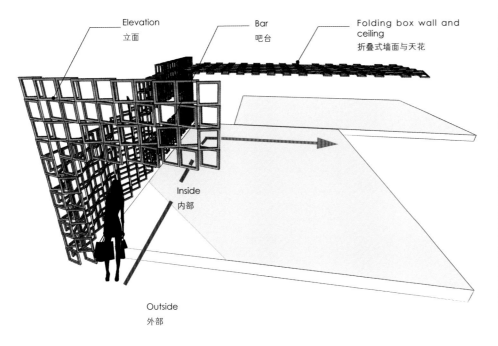

Elevation 立面
Bar 吧台
Folding box wall and ceiling 折叠式墙面与天花
Inside 内部
Outside 外部

Successful Bakery Design II 181

Boulanger Kaiti

日本福冈卡提面包店 Fukuoka, Japan
日本，福冈

Beauty born from simpility
源自简约之美

Basic information 基本信息

Design/ 设计：MOVEDESIGN
Designer/ 设计师：Mikio Sakamoto/ 坂本干雄
Renovation area/ 翻新面积：138.05 m²
Photography/ 摄影：Yousuke Harigane (Techni Staff)
Completion/ 完工时间：2014

Key materials 主要材料

Wood
木材

A bakery refurnished from a historical residential house – how to fit into the existing environment is the main challenge.

The bakery is renovated from an old Japanese house in Fukuoka, a residential district close to downtown. The existing house built of wood has impressive old wood walls and tile-roofing. As for the neighbouring houses, it could be seen that this house has a history and exists as a symbol of this town.

这家店面是从一座古老的民宅改建而成。设计要考虑的主要问题是如何使其融入原有的环境。

这座传统日式民宅位于福冈市，所在的住宅区离市中心很近。房子是木质结构，斑驳的木板墙和整洁的瓦片屋顶都极具特色，在周围建筑中显得独树一帜，已经成为这个城区的标志性形象了。

The owner's passion and his personal character percolate through this place and create 'Kaiti's view of the world' here.

设计营造出一个"卡提世界",店内处处渗透了卡提品牌特有的韵味。

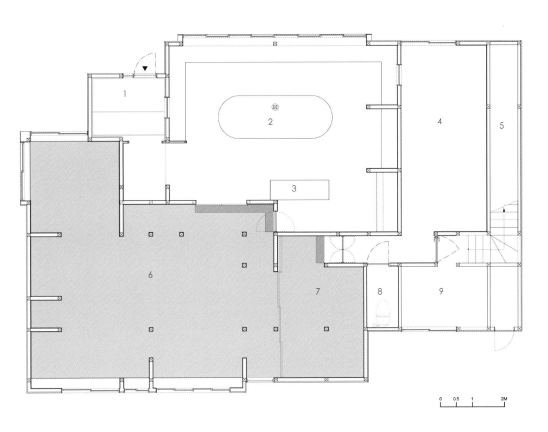

Floor Plan
1. Entrance
2. Bakery
3. Cashier
4. Stock
5. Garage
6. Main kitchen
7. Sub kitchen
8. W.C.
9. Staff room

平面图
1. 入口
2. 面包店
3. 收银台
4. 备货区
5. 车库
6. 主厨房
7. 次厨房
8. 卫生间
9. 员工休息室

Successful Bakery Design II 185

The owner's philosophy is 'to offer delicious quality bread everyday'. He does not permit compromise on anything, selecting ingredients and deciding how to make bread.

店主的信条是：每日供应高品质的美味面包。从选材到制作过程，任何细节，绝不妥协。

midi a midi

迷迪面包店 Kanagawa, Japan
日本，神奈川

Experiencing the new-concept bakery, just like enjoying books in the library
像在图书馆阅读一样，体验新概念面包店

Basic information 基本信息

Design/ 设计：Aki Hamada Architects + Kentaro Fujimoto
Area/ 面积：104.9 m²
Photography/ 摄影：Takumi Ota
Completion/ 完工时间：2014.12

Key materials 主要材料

Wood
木材

The series of activities to choose bread, take it out and eat it in the bakery is similar to the experience enjoying the book in the library. The designers wanted to create the new experience in the bakery by referring to the various activities in the library. There is the experience in the library that searching for the book is fun in itself.

挑选面包，将它拿出来，在面包店吃下去，这一系列活动与在图书馆中阅读图书的过程十分相似。设计师希望通过参考图书馆为面包店带来全新的体验。在图书馆里，寻找图书本身就是一件有趣的事情。

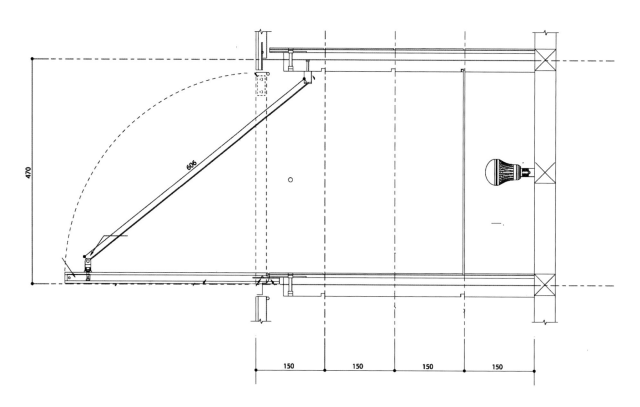

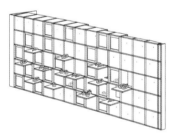 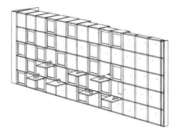 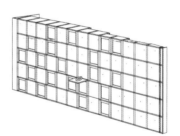 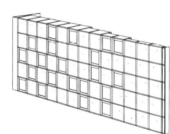

The designers created the new shelf system to provide an accidental discovery of the bread as our experience in the library. The shelves for the bread always change their density of the bread, because the bread is sold and replaced by new one. They wanted to make optimal display of the bread by opening or closing shelves. With the flexible and interactive system the clerk and customer can face bread through the shelves every time.

设计师打造了全新的货架系统,为我们提供了类似于图书馆体验的面包发现之旅。货架上的面包总在不断变化,因为有的面包被卖掉了,又补充了新的面包。设计师希望通过货架的开放和关闭呈现最佳的展示效果。这是一个灵活的互动系统,店员和顾客可以通过货架直接面对面包。

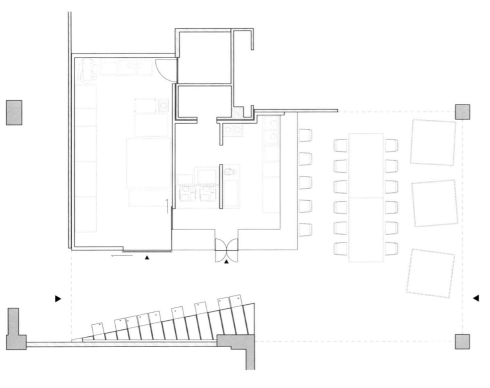

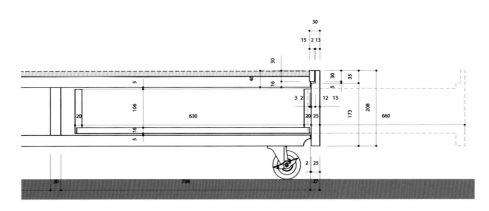

There are three types of seating: counter seat, table seat, TATAMI seat. They enhance selectivity among the small spaces.

A mobile TATAMI is used by connecting together with two or three and let out to outside. It is possible to customise in order to use it by the various formations.

店内共有三种座位，吧台座、餐桌座和榻榻米座，它们为小空间提供了更多的选择。

移动的榻榻米家具可以合并起来，也可以移到室外，根据不同的需求，它们可以实现多种造型。

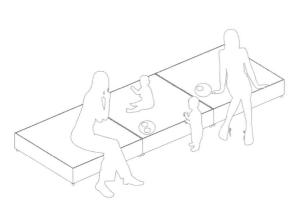
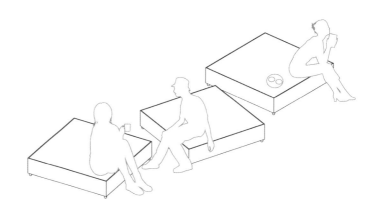

T by Luxbite

T 甜点屋 Victoria, Australia
澳大利亚，维多利亚

Best things in small package
"小"空间内的"大"精彩

Basic information 基本信息

Design/ 设计 : WALA (Weian Lim Architects)
Area/ 面积 : 19 m²
Photography/ 摄影 : Fraser Marsden (www.fraermarsden.com)
Completion/ 完工时间 : 2014.11

Key materials 主要材料

Steel
钢材

The design of the new addition to the Luxbite family, patisserie 'T by Luxbite', aims to show that the best things can come in small packages, just like their pastries within. The overall design focuses on the presentation and quality of the shop's unique products; complementing them without overshadowing them. (www.tbyluxbite.com.au)

卢克西比特家族所经营的"T 甜点"的设计目标是展现小空间也能有大精彩，就像他们所销售的甜点一样。空间的整体设计聚焦于展示和独特产品的品质，设计不能喧宾夺主，而要锦上添花。

The owners required five separate purpose-built fridges to fit within the shop's footprint. Inspired by cafes in Japan (which the owners adore!) where creative use of small spaces permeates through the design, a clever layout ensures that all kitchen equipments were well located for staff to work comfortably. The result is also more-than-generous amount of storage throughout. The minimalistic approach continues with a simple material palette, coupling with straight geometries to achieve a regulated sense of order.

Despite the small footprint, the floor spaces have designated front and back-of-house areas where tarts are made in-house. However, with no space to conceal kitchen activities, food preparation work becomes theatre. To minimise clutter, storage joinery was selectively designed to squeeze every last inch of useable space. Built-out veneered walls, overhead cupboards and counters all conceal access to services and storage.

店主要求在店内摆放5台冰箱。受到日式咖啡店(店主十分热爱日式咖啡店)巧妙利用小空间的启发,巧妙的布局保证了所有厨房设备各得其所,员工工作舒适便利。设计还在店内各处提供了充足的储藏空间。极简设计对应了简单的材料搭配,整齐笔直的几何线条给人以秩序感。

尽管空间面积很小,店铺还是严格划分了前台和后台区域。由于无法隐藏厨房活动,食品准备工作变成了一种表演。为了更整洁,特别设计的储藏柜尽力挤出每一寸空间。外置式镶面墙、吊柜和柜台巧妙地将服务和储藏通道隐藏了起来。

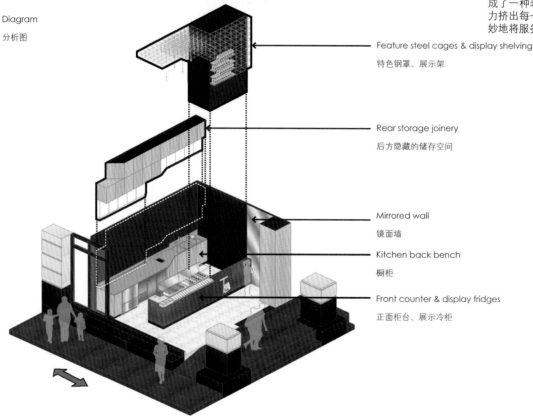

Diagram
分析图

Feature steel cages & display shelving
特色钢罩、展示架

Rear storage joinery
后方隐藏的储存空间

Mirrored wall
镜面墙

Kitchen back bench
橱柜

Front counter & display fridges
正面柜台、展示冷柜

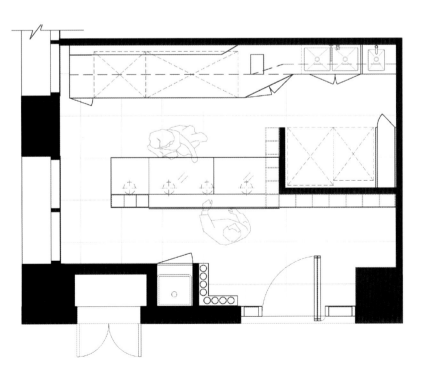

The shop's name is derived from their sale pastry of choice, tarts. Similarly, the meticulous interior design draws inspiration from the intricately crafted tarts. The 19-square-metre floor corner location along Flinders Lane required a stronger emphasis on shopfront prominence and maximising the use of all available internal space.

店铺的名字来自于他们主营的甜点——水果馅饼（Tart）。同样的，精致的室内设计也从水果馅饼的复杂工艺中获得了灵感。19平方米的街角空间需要视觉感更强烈的店面外观和对室内空间的最大化利用。

Overhead steel 'cages' wrap from the ceiling and walls to conjure a sense of layered depth. This feature works in tandem with dramatic lighting (via striplights and suspended globes) to lend a sense of occasion to the space. An adjacent full-height mirror serves not only to create the illusion of an enlarged space, but to also mirror the steel cages to display the letter 'T', thereby reinforcing the shop's brand.

Clear lines, geometric shapes and warm textures collectively promote a more subdued backdrop for the products to shine when on display. This consistent palette also ensured all elements of the shop sit harmoniously with each other.

吊顶"铁笼"将天花板和墙壁包覆起来，突出了层次感。这一设计与夸张的灯光共同赋予了空间一种戏剧感。旁边的落地镜不仅能在视觉上扩大空间，还能映出铁笼的T字型，突出店铺的品牌形象。

简洁的线条、几何图形和温馨的材质为产品提供了柔和的背景，使它们散发出光辉。这种连贯的设计还保证了店内各种元素的和谐统一。

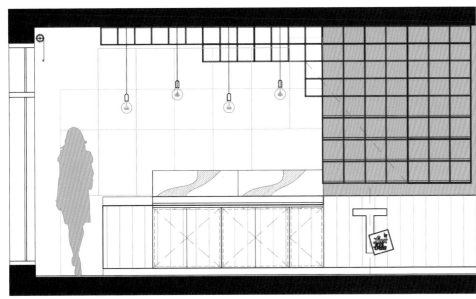

Style Bakery

风尚面包店 Gunma, Japan
日本，群马

Bread is the protagonist in the simple Japanese style bakery
面包是简约日式风尚面包店里的主角

Basic information 基本信息

Design/ 设计 : SNARK
Design team/ 设计团队 : Sunao Koase, Yu yamada
Area/ 面积 : 280 m²
Photography/ 摄影 : Ippei Shinzawa
Completion/ 完工时间 : 2014

Key materials 主要材料

Paint (exterior); tile, paint (wall); planking, paint (ceiling)
涂料（外部）；瓷砖、涂料（墙面）；木板、涂料（天花板）

In Kiryu, Gunma, Japan, Style Bakery was opened as the first prototype shop with a future view of store development. Renovating the long-established bakery operating since 1930, the shop aims to be a new type of enterprise, incorporating the features of both locally owned and globally franchised stores.

在日本群马县的桐生市，这家风尚面包店是未来连锁店的首家原型店，起到了奠定基础的作用。这家面包店创立于1930年，店铺的翻修目标是打造一种全新的店铺形式，既富有当地特色，又具有国际化加盟店的特征。

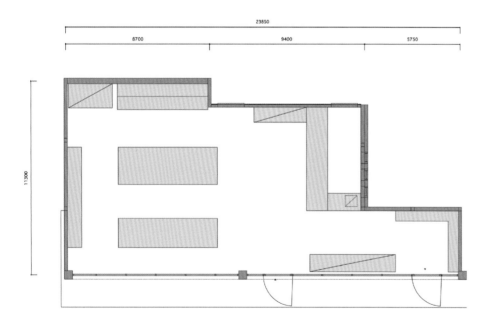

A light grey, which is the concept colour of the brand, is used in tiles, lights, steel frames, and a dirt floor. Variations of materials used balance the colour tint within the space, bringing out patisseries. Using the light grey as the concept colour of the brand, the shop holds the sense of unity, and bright and clean atmosphere at the same time, together with woods, soft lighting, and natural sun light from the windows. Solid oaks are used for the top panel so as to enhance the warmth of the bread.

Maximising the height and the width of the windows and selecting a slim frame, it will bring brightness and freshness to the room. The space right next to the shelf is refrigerated so that foods like sandwiches can be displayed.

Eat-in counter located next to the cashier, is made with the same solid oak as the top panel, and uses the marine lamps.

作为品牌的主题色彩，浅灰色被应用在店铺的各个角落：瓷砖、灯具、钢架、地面。为了平衡室内空间的色调，设计师选用了各种金属材料，以烘托出作为主角的糕点。以浅灰色为主题色，店铺具有一种统一感，与木材、柔和的灯光和阳光相融合，显得明亮而干净。橡木台面的设计突出了面包的温暖感。

把窗口的高度和宽度最大化，同时选择较细的窗框，为室内带来了明亮和清新的感觉。紧靠货架的冷藏区可以摆放一些三明治类的食物。

堂食区紧邻收银台，其台面同样采用橡木，头顶采用船用灯进行照明。

Fluorescent lamps are embedded into the ceiling, which makes the space look clean. Larger furniture in the centre and the spot shade above the cashier build vertical internal.

Created line of flows, with the simplified designs and arrangements of furniture and lined LED lights equipped down under the parietal shelves, will enable the customers to choose and carry the product in comfort.

The signboard, which will be the representation of the shop, is neon lamp, and the entrance / exit signs are located on both sides. Additional attentions were paid for children and the elderly to secure their paths, by sloping the way from the parking.

荧光灯嵌在天花板上,使空间看起来十分简洁。摆放在中央的大家具和收银台上方的吊灯在空间内建立了垂直感。

极简设计、家具的有序摆放、LED照明以及合理的动线设计,让顾客可以舒服地选择和购买商品。

店铺的招牌采用霓虹灯,旁边设有出入标志。为了方便儿童和老年人的出入,门口采用坡道设计,安全便利。

So Milaky

索·米拉其面包店 Shaoxing, China 中国，绍兴

Healthy concept + exotic atmosphere + warm-toned bread
健康质朴的理念 + 异域情怀的环境 + 暖色美味的面包

Basic information 基本信息

Design/ 设计 : Associate Interior Design/ 杭州意关联设计事务所
Design team/ 设计团队 : Ge Jianwei, Ye Zengmin, Ma Junhui, Wang Sheng/ 葛建伟、叶增敏、马军辉、王盛
Area/ 面积 :148 m²
Completion/ 完工时间 : 2014

Key materials 主要材料

Reused plank, steel plate, glass, white tile, cement paint
老木板、钢板、玻璃、白砖、水泥漆

Located at Yintai Mall in Shaoxing, China, So Milaky Bakery has a total area of 148 square metres. With this limited space, the designers created a composite space that combines production, sale and dining together. In accordance to the owner's organic food concept and pursuit of natural, healthy and simple life, solid wood is extensively used throughout the interior space. Besides, the effective product display also manifests the design's practicality.

位于绍兴金帝银泰城的索·米拉其面包店总面积148平方米，设计师充分利用有限的面积构建起一个加工、售卖、就餐的综合空间；为了迎合业主销售有机食品所追求的天然、健康、质朴的理念，设计师在此大量使用了实木材料，同时高效率的商品展示能力也充分体现了本案的功能性和实用性。

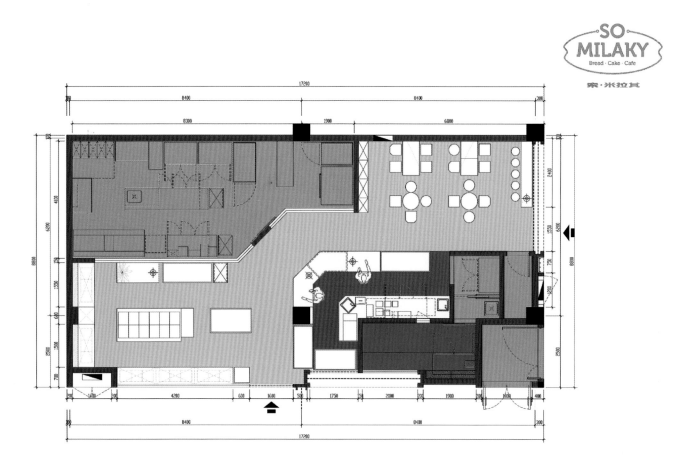

Bakery design is different from other dining interior design types. In this case, food has the opportunity to be displayed elaborately. The collocation between vintage exotic atmosphere and warm-toned bread creates a warm effect, as well as emphasises the colour feature of bread.

面包店在餐饮类别中比较特别，食物有机会作为主体被精心展示。考虑通过复古有异域情怀的环境和暖色调的面包搭配，渲染出温馨的效果，同时突出了面包的色彩特征。

Dining area uses the same colour palette as bread display area. The books, decorations and plants on the wall cabinets create a home-like warm atmosphere.

就餐区的设计在色彩运用上与面包展示区相呼应，墙壁上的小橱柜上摆放的书籍、装饰以及绿色小植营造了居家般的温馨氛围。

Serrajòrdia

塞拉约迪亚面包店 Barcelona, Spain
西班牙，巴塞罗那

A traditional business with an industrial and vintage touch
加入工业风格和复古气息的传统面包店

Basic information 基本信息

Design/ 设计 : AM Asociados
Area/ 面积 : 174 m²
Photography/ 摄影 : Marta Pons

Key materials 主要材料

Distressed wood (floor, counter)
black iron
做旧木板（地面、台面）
黑色铸铁

The interior design studio AM Asociados has carried out the design and implementation of Serrajòrdia, an emblematic bakery in the town of Sant Cugat del Valles near Barcelona.

The space is designed with a dual role in mind: a sales area with coffee shop and a delicatessen. The objective was to capture the maximum natural light possible in order to enhance the product and provide light to the whole area.

塞拉约迪亚面包店位于巴塞罗那附近的圣库加特德瓦利斯镇，是当地一家标志性的面包店，AM Asociados室内设计工作室为其进行了全套的设计。

面包店的空间具有双重身份：既是一家咖啡馆，又是一家烘焙食品店。设计的目标是最大限度地利用自然采光来美化产品，为整个室内区域带来光亮。

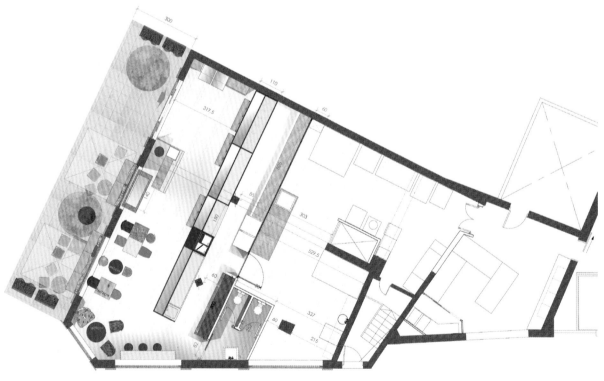

The industrial touch has been achieved through several factors. Firstly, the integration of the workrooms with the sales area by using iron fences and glass. Thus everything is within the view of the clients, providing the premises with a feeling of great spaciousness.

Secondly, all the installations are visible, the air ducts, water pipes, electricity, fire, structural beams and the wood and iron fittings, all lit with lights specially chosen to maintain excellent luminosity throughout the whole day.

设计利用几种元素实现了工业氛围。首先，工作室和销售区通过铁栅栏和玻璃连接起来，让一切尽收顾客的眼底，让空间显得十分宽敞明亮。其次，所有装置都裸露在外，包括通风管道、水管、电线、防火设施、结构梁、木工装饰和五金件等。精选的照明设施让空间全天都能保持十分明亮。

The furniture is made of fine materials inspired by a vintage style combined with white roughly textured walls with a brick backdrop in its natural colour.

The exhibition and sale furniture has been designed and manufactured by AM Asociados. Some elements that stand out are: the distressed wood on the floor and on the counter, as well as the black-iron structures. Another basic element is the large display rack, made of distressed solid pine.

Furniture brands Francisco Segarra and Merc & Co sign the remaining furniture. The lamps were purchased in Factory 20 and Fins de Siegles and combined to create a welcoming atmosphere.

家具全部采用精致材料制成，具有复古风情，与纹理粗糙的白色墙壁和砖墙背景完美融合。

展示货架和货柜全部由 AM Asociados 设计并制造。其中较为出色的元素包括：地面和台面上的做旧的木板、黑铁结构等。巨大的展示架由松木制成，同样引人注目。

店内的其他家具来自 Francisco Segarra 和 Merc & Co 两个品牌。灯具采购自 Factory 20 和 Fins de Siegles，营造出一种好客的氛围。

Chocolateria Brescó

布雷西科巧克力烘焙坊

Barcelona, Spain
西班牙，巴塞罗那

A product that requires a long manipulation process, and a unique and beautiful building

一种需要长时间制作的产品，一座独特而美丽的建筑

Basic information 基本信息

Design/ 设计：desafrà
Area/ 面积：160.27 m²
Completion/ 完工时间：2013

Key materials 主要材料

Wooden wall
木墙

Here, you find a furniture display that, through a video synchronised with lights, make appear and disappear objects that allows to explain the cacao history and the chocolate elaboration process. Finally, the last stop is within the toilets. On the walls, you can find texts and images about Casa Calvet, as well as Gaudí's biography.

烘焙坊的墙壁上有一个显示屏，向人们解释了可可豆的历史以及巧克力制作的过程。洗手间的墙面上有卡佛之家的介绍文字和图像，还有高迪的自传。

Due to the fact that currently there are 350 references of the product, and that some need to be maintained in a cold area and others not, it needed a big length of shelves and refrigerated displays. The programme distribution places the area of shelves and refrigerated displays closer to the street access. Later on, we will find the area of displays and shop, connected with the tasting place with bar. At the end of the space, toilettes are located.

All intervention, such as furniture, lighting, installations... are located not higher than mid height. Lighting extols ceiling architecture, sometimes forgotten. Driven by the need to create a warehouse, the original wooden wall has been reproduced with glass windows at the end of the space. Without this intervention, this modernist element would be hidden to the public.

目前，烘焙坊出品的产品有350种，其中一些需要冷藏保存，而另一些不需要。因此，它需要一个长长的货柜和冷藏展示区。设计师让货柜和冷藏柜的位置靠近街道。再往里走，品尝区把展示区和商店连接了起来。而空间的最里面则是洗手间。

家具、照明、摆设等改造设施的高度全部不高于空间的中间高度。为了营造一种仓库的效果，设计在空间的最后重新复制了木墙，并添加了窗户。如果没有这一设计，这件现代艺术品将会消失在人们的视线中。

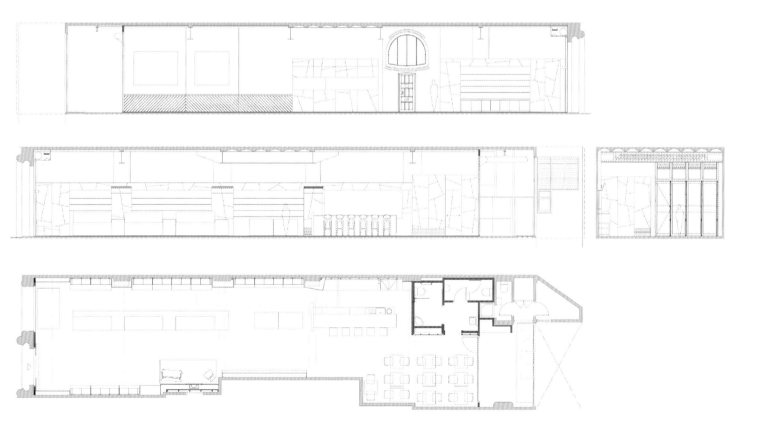

Due to fire prevention rules, it has been requested to place a new wall in front of the door and window that originally connected the space with the lobby of the building. But this intervention has been done carefully reproducing this original part in the new wall surface, to allow to view the work and spirit of Gaudí's architecture. In specific places has been reproduced the original door support, disappeared long ago.

The new façade carpentry has been hidden, to provide a clean view of the original stone of the arcades. Once entering the premises, you find the ornamental cover with the reproduction of the window, door, furniture, peephole... a Gaudí's corner. The bench, the chair and the peephole were all designed by Gaudí himself, especially for the Casa Calvet.

防火规定要求在连接空间与大厅之间的门窗前方新修建一面墙。设计师通过巧妙的设计，在新墙面上重现了原始的纹理，体现了高迪的建筑精神，并且重新制作了门框。为了营造原始石拱门的简洁感，新的外墙木工设计被隐藏了起来。

走进烘焙坊，你会发现无论是门窗、家具乃至门镜，都体现着高迪的风格。长椅、座椅和门镜全部由高迪亲自为卡佛之家设计。

IL LAGO Bakery & Wine Shop in MVL Hotel Kintex

韩国国际会展中心 MVL 酒店湖泊烘焙坊与酒廊   Goyang, Korea 韩国，高阳

Cultural and emotional values can be preserved through design - conversation with the hotel

设计留存文化与情感价值——与酒店之间的对话

Basic information 基本信息

Design/设计：Daemyungresort Bae Yun Joon/ www.daemyungresort.com/
Design Bono. Sung Jin, Jang/ www.designbono.com
Area/面积：150 m²
Photography/摄影：Pyo-Jun Lee
Completion/完工时间：2013.4

Key materials 主要材料

Polishing tile (floor); white urethane paint, corrosion paint (wall); vinyl paint (ceiling)
抛光瓷砖（地面）；白色聚氨酯涂料、防蚀涂料（墙面）；乙烯基涂料（天花板）

The image motivation of 'IL LAGO' is reminded wandering sailboat which follows daylight through layered arches in a culvert in Europe where canal systems have been developed.

These days, through a lot of shops that are benchmarked European designs and vintage concept shops, people could marvel for the neat and classic design output. However, actually, on the site, it is really difficult to bring sensibility of history. In Europe, you can see that they do remodelling culverts of Thames River to pubs protecting historical structures of Thames riverside and the design shows cultural and emotional values. Thus, the designers need to run back their historical space again and approach with new viewpoint.

湖泊烘焙坊的形象好像一艘漂流的帆船，它在黎明穿过欧洲城市运河留下的层层涵洞，悠然自得。

不同于典型的欧洲设计和复古风格的店铺，人们会惊叹于本店设计的简洁优雅。事实上，在店铺所在地，很难塑造一种历史感。在欧洲，开发者常常将运河的涵洞改造成酒吧，同时又保护了河畔的历史结构，使设计彰显出文化与情感价值。因此，他们也不断回顾自己的历史空间，并且以全新的观点进行处理。

This project began by using a blank space at MVL hotel which is a recently established Hotel next to the Ilsan Kintex in Seoul as a bakery and wine shop. In accordance with requirements of the client, the designers tried to overcome the limit of the location. In addition, they created sculptural structure to develop the lobby design and then the sculptures are connected from the shop to the entrance area, and it could make the shop well perceived from the entrance. Furthermore, the designers attempted to create a huge sculptural space by using the sculptural cupboard which looks like a bottom of column and arch object.

The designers planned the lobby and the façade of the shop as one of sculptures on the ground with sculptural elements by using an open plan, openings and high ceiling. Thus, it could make people get nice mumps and then it could emboss awareness of the shop.

该项目位于韩国国际会展中心附近MVL酒店里的一块空白区域。为了满足客户要求，设计师力求克服地点的限制。此外，设计师还创造了富有雕塑感的结构来进行大堂设计。雕塑结构将店面与酒店大门连接起来，使店面更加引人注目。设计师还尝试设计了一个富有雕塑感的橱柜，使其看起来像是立柱或拱门的底部的一部分。

设计师将酒店大堂和店面设计成一个整体雕塑，采用开放式布局、入口和挑高的天花板。这种设计既能让人心情愉悦，又能为店铺吸引更多的注意力。

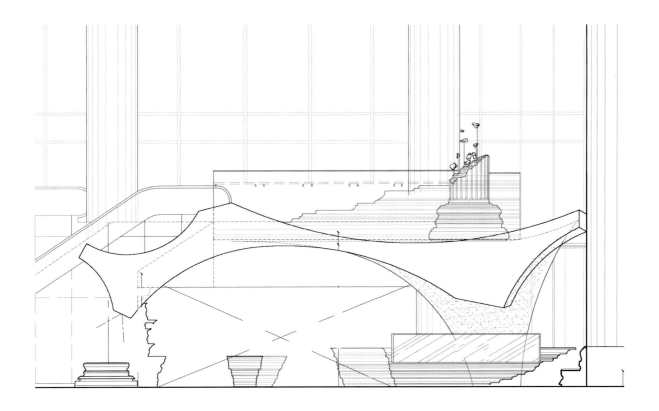

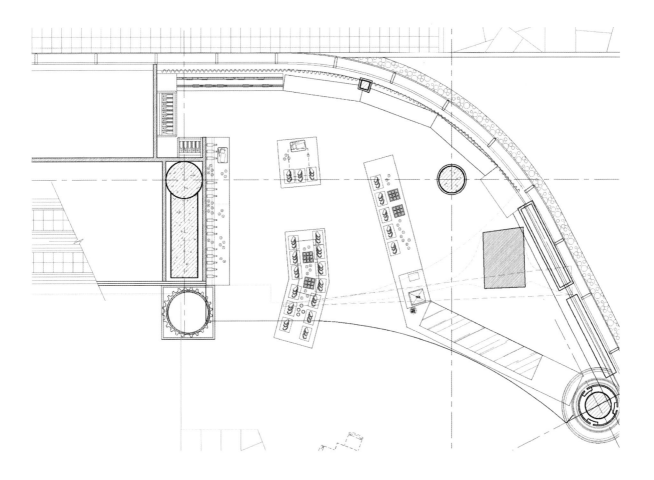

Arch shape structure and classical layered molding is harmonised with façade reinterpretation. The arch, which could be seen frequently through a space having historical background in Europe, has been used as an architectural skill for depth of repeated shape and daylight into the space. Furthermore, the arch made a cave-shaped space in the large lobby, and also it tried to make independent spaces in accordance with characteristics of the bakery and wine shop.

Especially, as an element of space, arch structure has been designed by line and face and column object and molding which are layered on the surface extremely emphasise the detail. In addition, it makes an independent structure by resisting and adapting the curve of the arch and the harmony of the massive object, curved line, horizontal line and faces, makes dynamic and live space. Thus, the design elements have built unique, luxurious and emblematic space of MVL Hotel as a deluxe hotel.

拱形结构和经典的层叠嵌线与店面设计完美契合。作为一种建筑技巧,在欧洲历史建筑中常见的拱门能展现重复空间的层次感,为内部引入自然光。此外,拱门在大堂内形成了一个深邃的洞穴空间,让独立的空间与烘焙坊和酒廊的风格保持一致。

作为空间元素,拱形结构采用点面设计,而立柱和嵌线在面上的层叠凸显了细节效果。拱形、弧线、水平面和平面的融合形成了一个充满了动感与活力的空间。所有设计元素共同营造出一个独特、奢华、具有象征意义的空间,凸显了MVL酒店作为奢华酒店的地位。

Gail's Bakery, Chelsea

英国伦敦切尔西盖尔面包店 London, UK
英国，伦敦

Aiming to create a backdrop full of aesthetic for the bakery produce
为面包打造充满艺术气息的背景

Basic information 基本信息

Design/ 设计：moreno:masey
Photography/ 摄影：Richard Lewisohn
Completion/ 完工时间：2013

Key materials 主要材料

brick work (floor), marble (counter)
裸露地砖（地面）、大理石（台面）

Located on London's sophisticated King's Road, this listed corner building contained a simply beautiful façade whose embellished windows were reminiscent of an old apothecary. In keeping with this aesthetic, a subtle palette of materials was selected to create backdrop for the display of the bakery produce.

这家面包店位于伦敦古老的国王路上，所在建筑历史悠久。店面外观简洁美观，窗户看上去有点像那种古老的药店。设计延续了这种风格，精心选择所用材料，为琳琅满目的面包营造了完美的背景。

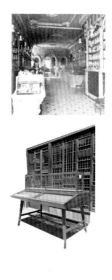

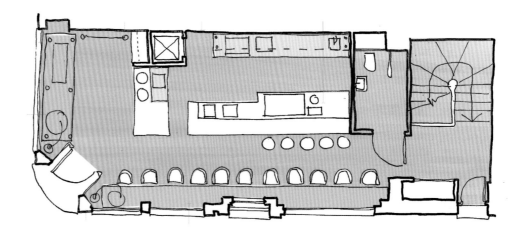

The qualities of the existing rough brickwork were enhanced by white washing and contrasting their texture against the smooth refined marble of the countertops. A restrained palette of interior finishes added distinctive touches of colour including warm timber, soft indulgent leather banquettes and reclaimed furniture.

Tables and pedestals were selected for the intricacy of their iron bases set against rows of distinctive cone-shaped lights with slender cabling. Carefully detailed joinery acknowledges the craftsmanship of the original shopfront design.

原有的粗糙砌砖表面喷涂了白色油漆，凸显了材料的粗糙质感，与光滑的大理石柜台形成鲜明对照。所用装饰材料种类并不多，但别具一格，为环境增添了几抹色彩，包括暖色的地板、柔软的皮椅以及回收利用的家具。桌椅都是铁质的，简洁大方，搭配一排排柱状吊灯。精致的工艺凸显了店面原有的高档品质。

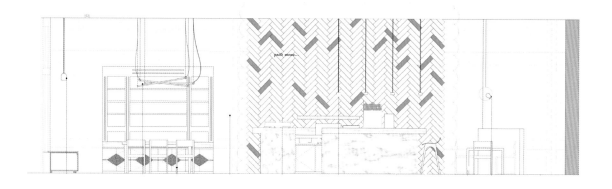

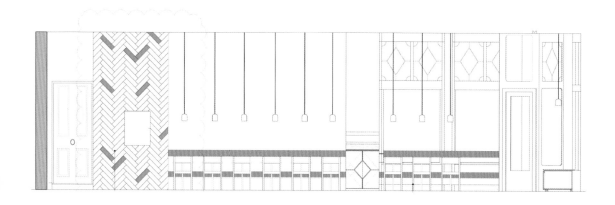

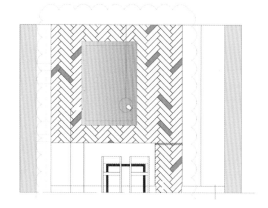
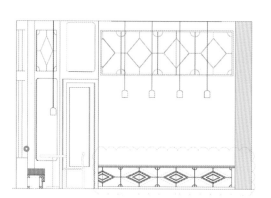

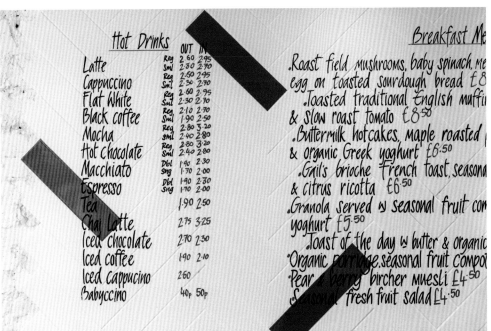

Hot Drinks		OUT	IN
Latte	Reg	2.60	2.95
	Sml	2.30	2.70
Cappuccino	Reg	2.50	2.95
	Sml	2.30	2.70
Flat White	Reg	2.60	2.95
	Sml	2.30	2.70
Black coffee	Reg	2.10	2.70
	Sml	1.90	2.50
Mocha	Reg	2.80	3.20
	Sml	2.40	2.80
Hot chocolate	Reg	2.80	3.20
	Sml	2.40	2.80
Macchiato	Dbl	1.90	2.30
	Sng	1.70	2.00
Espresso	Dbl	1.90	2.30
	Sng	1.70	2.00
Tea		1.90	2.50
Chai Latte		2.75	3.25
Iced chocolate		2.70	2.30
Iced coffee		1.90	2.10
Iced cappucino		2.60	
Babyccino		40p	50p

Breakfast Menu

- Roast field mushrooms, baby spinach, me... egg on toasted sourdough bread £8...
- Toasted traditional English muffin... & slow roast tomato £8.50
- Buttermilk hotcakes, maple roasted... & organic Greek yoghurt £6.50
- Gail's brioche French toast, seasonal... & citrus ricotta £6.50
- Granola served w seasonal fruit com... yoghurt £5.50
- Toast of the day w butter & organic...
- Organic porridge, seasonal fruit compote...
- Pear & berry bircher muesli £4.50
- Seasonal fresh fruit salad £4.50

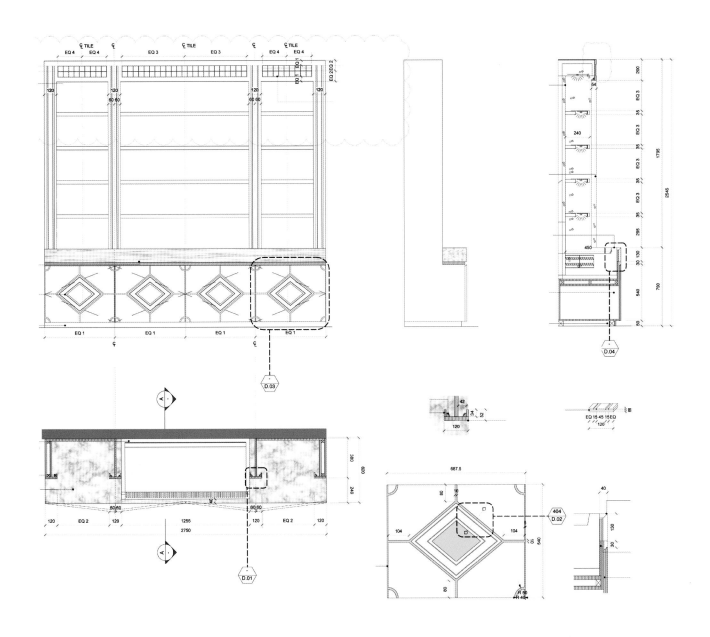

Open style cabinets compliment the detail of the joinery and allow the prominent presentation of the artisan produce. The cabinets are highlighted by subtle lighting strips while Gail's signature red was introduced into the herringbone pattern, tiled wall features as well as the bespoke joinery ensuring optimisation of the brand identity.

开放式陈列柜凸显了细木工的精致细节，同时也衬托了里面陈列的商品。陈列柜用灯带照明，"盖尔"的店名标识隐藏于图案之中，突出了专属定制的品位，彰显了品牌形象。

Index 索引

Airhouse Design Office
http://www.airhouse.jp/

Aki Hamada Architects + Kentaro Fujimoto
http://aki-hamada.com/

AM Asociados
http://amasociados.com/

Anagrama
http://www.anagrama.ro/

Andrea Langhi Design
www.andrealanghi.it

Anne Sophie Goneau
www.asgoneaudesign.com

Architect David Guerra
http://davidguerra.com.br/

Arnau Vergés Tejero – arnau estudi d'arquitectura
http://www.arnauestudi.cat/

Associate Interior Design
http://weibo.com/u/5395008398?from=feed&loc=nickname&is_all=1

Atelier Moderno
www.ateliermoderno.com

Backbone Branding
http://www.backbonebranding.com/

Collidaniel architetto – Rome
www.collidaniela.com

Crosby Studios
http://www.crosby-studios.com/

Daemyungresort Bae Yun Joon
www.daemyungresort.com/

desafrà
www.pepbolta.com

David Pinilla
www.pinillainside.com

Design Bono. Sung Jin, Jang
www.designbono.com

Designliga
http://en.designliga.com/

Er Heyong Space Design Group

ideoarquitectura
http://www.espacioideo.com/

Ico Design
http://www.icodesign.com/

JC Architecture
www.johnnyisborn.com

Kozyrniy Design
www.k-design.com.ua

LINK arkitektur
www.linkarkitektur.no

LUKSTUDIO
http://www.lukstudiodesign.com/

Memo & Moi Brand Consultants
www.facebook.com/memoymoi

mode:lina architekci architecture studio
http://modelina-architekci.com/

moodley brand identity, grafisches Büro
http://www.moodley.at/

moreno:masey
http://www.morenomasey.com/

MOVEDESIGN
http://movedesign.jp/

NAN Architects
www.nancontract.com

Nordic Bros. Design Community
http://www.nordicbrosdesign.com/

Savvy Studio
http://savvy-studio.net/

Slava Balbek, Nadya Chabanny
http://www.2bua.com

SNARK
http://www.snark.cc/

Synthesis Quatro / Slavica Djokovic /Aleksandar Osei-Lartey
http://www.slavicalazic.com/

WALA (Weian Lim Architects)
http://www.wa-la.net/

Appendix

附录一：烘焙坊选址

如今，烘焙食品已成为了很多消费者的挚爱，烘焙店的发展趋势也越来越好。烘焙店除了选择项目品牌、解决资金和技术等是非常重要的前期准备之外，在确定项目品牌之后的开店选址可以说是在前期计划中，最能影响后期销售和利润的步骤。因此，烘焙店选址一定要经过多方面的考察，遵循选址原则，做好充分的调查和对比，找到最适合的位置，为之后的经营打下良好的基础，从而带来可观的经济收益。

影响选址因素

1. 周围环境

环境的好坏包含两层含义。其一是指店铺周围环境状况，比如卫生情况，是否有污染性的工厂等。另一层含义指店铺所处位置繁华程度。一般来说，店铺若处在车站附近、商业区域人口密度高的地区或同行集中的一条街上，这类开店环境相对具有比较大的优势。另外，交叉路口、拐角的位置一般较好，而坡路上、偏僻角落或者楼屋高的地方则位置欠佳。

2. 交通便利性

交通是否便利对烘焙坊店铺的销售有着很大的影响。顾客到店后，停车是否方便；货物运输是否方便；从其他地段到店乘车是否方便等。如公共交通路线很多、街道很宽敞、或停车位置足够使用，则生意自然就会多了。

烘焙坊选址在幽静的小巷内，带有独立的庭院，店主追求清新、优雅的生活。目标顾客群体主要为喜欢安静、乐于消磨时光的人

烘焙坊选址在商业街上,来往的人群是主要的顾客流,因此其在装修风格上也以实用性为主

烘焙坊选址在酒店大堂内,酒店客人成为主要的顾客群体,独特的设计格外引人注目

3. 商圈背景与住户性质

调查商圈内各类别所占的比例。在办公楼区,则上下班时间生意会相对较好;在住宅区任何时间生意都会不错,但同时房租可能会高出很多;若店铺附近有机关学校等大团体设立,则会带来固定之消费人群,但放假时,生意就会很差。

商圈内住户的基本背景,如收入、教育、职业、家庭大小、宗教信仰、年龄等都是影响购买行为的因素。夫妇两人教育水平愈高,则夫妇两人同时就业比例愈大,且在外吃饭或购买面包之机会就愈大。住户收入愈高,消费能力愈强,则生意会愈好。

4. 竞争程度

商圈内相同类型的店面愈多,则生意愈竞争。面包是一般食品,与其相关的食品店之中,西式速食店、餐厅、夜市、路边摊等都是烘焙坊的替代店,所以这类店面愈少,生意会愈好。

5. 未来发展

烘焙坊店址所处的区域未来发展愈快,对生意的帮助就愈大。如果附近有马路或地铁等的兴建,都会为以后带来更多的人潮流量。

小建议

1. 开烘焙店选址的时候,投资者一定要注意自身定位。对于自己的财务水平有正确的评估。如果资金充裕,那么可以尽量选择高档商业区或者高档社区附近,在中高水平消费人群众多,客流量密集的地方。高档的烘焙店只要环境和食品相匹配,然后再加上优质的服务一定会让收益增加许多,还能"拉拢"众多忠实的消费者。当然必须要牢记一点,这样的地方,租金和人员服务成本都会大大增加,风险也会增加。

2. 如果投资者资金相对紧张,可以选择客流量大的街区或者社区,但要综合附近的居民消费水平来综合考虑产品的定价,制定符合地区性质的销售计划。例如,学校旁边一定要考虑寒暑假期的特定时间,把这段时间的花费算到成本里。无论如何,学校附近的烘焙店还是相对具有一定的优势,因为客流量大,而烘焙食品通常也是学生群体非常忠实的持续消费者。

附录二：烘焙坊设备工具

烘焙坊所需基本设备及工具主要包括烘焙模具器材类和电器机械类。

烘焙模具器材类

序号	名称	备注
1	量杯	通常一量杯标准液体容量240毫升。量杯材质有铝、塑胶等
2	量匙	用于称取少量材料，一套通常由4支组成，分1大匙、一茶匙、1/2茶匙和1/4茶匙（1大匙=3茶匙）
3	打蛋盆	通常分为大、中、小三种，常用直径分别为38、32、26厘米
4	面粉筛	一般使用不锈钢筛网，可兼做过滤使用，筛目一般为30目
5	橡皮刮刀	用于刮净黏附在搅拌缸或打蛋盆中的材料，也可用于材料调拌，分为多个型号
6	刮匙	用于包馅料，又称包料匙
7	挤花嘴	用于挤出各式霜式图样的金属模型管
8	挤花袋	用于霜式材料挤出之用，通常有帆布、塑胶及尼龙材质三种
9	吐司模型	专供烘焙吐司面包使用，通常根据面团重量划分规格，常用的有20两和12两两种
10	蛋糕模型	一般为铝制或不锈钢材质，规格形状各不相同
11	饼干模型	通常为不锈钢材质，形状各异
12	烤盘	通常为黑铁皮金属材质，用于食品烘焙之用。目前，主要有硅胶不沾烤盘、玻璃纤维烤垫等
13	面包出炉架	面包出炉后，冷却之用
14	切片机	供吐司面包切片之用

电器机械类

序号	名称	备注
1	电动打蛋器	又可称为手持搅拌器，通常有大、中、小三种规格，依照形状分为螺旋形和直形两种
2	电子称	通常使用的最大称量为2200克，其误差较小
3	烤箱	通常使用的有电气烤箱、瓦斯烤箱、蒸汽烤箱，其型式包括旋转式、坠道式、箱式和输送式等
4	搅拌机	一般分为横式和立式。横式多用于大型面包厂；立式通常附有三种拌打器，桨状、钩状和网状，一般烘焙店均可使用。注意立式搅拌量分为一贯和四贯两种，使用时最大搅拌量不宜超过其2/3。（目前，推荐一种新式面团搅拌机——直立式欧洲型搅拌机，螺旋式搅拌器，搅拌均匀、速度稳定）
5	发酵箱	面团发酵使用，如今电脑自动控制温度和时间的新型发酵箱比较方便实用
6	冰柜	透明冰柜或蛋糕陈列柜
7	微波炉	

除上述器材之外，还需烘焙专用油纸及油布、一整套刀具及烤箱纸。另外，根据烘焙店规模以及经营产品类型，应自行增添所需设备及工具。

图书在版编目（CIP）数据

烘焙坊. Ⅱ /（美）泰勒·鲁宾逊编；李婵译. — 沈阳：辽宁科学技术出版社, 2016.8
　　ISBN 978-7-5381-9848-5

Ⅰ. ①烘… Ⅱ. ①泰… ②李… Ⅲ. ①糕点—饮食业—服务建筑—室内装饰设计 Ⅳ. ① TU247.3

中国版本图书馆 CIP 数据核字 (2016) 第 142858 号

出版发行：辽宁科学技术出版社
　　　　　（地址：沈阳市和平区十一纬路 25 号 邮编：110003）
印　刷　者：恒美印务（广州）有限公司
经　销　者：各地新华书店
幅面尺寸：215 mm×285 mm
印　　张：15 1/4
插　　页：4
字　　数：50 千字
出版时间：2016 年 8 月第 1 版
印刷时间：2016 年 8 月第 1 次印刷
责任编辑：鄢　格
封面设计：何　萍
版式设计：何　萍
责任校对：周　文

书号：ISBN 978-7-5381-9848-5
定价：268.00 元

联系电话：024-23284360
邮购热线：024-23284502
http://www.lnkj.com.cn